White Poplar, Black Locust

White Poplar, Black Locust

Louise Wagenknecht

Oregon State University Press Corvallis

Acknowledgment for the use of previously published material appears on pages vii and viii.

Library of Congress Control Number: 2021948726. Cataloging-in-publication data is available from the Library of Congress.

∞ This paper meets the requirements of ANSI/NISO Z39.48-1992 (Permanence of Paper).

ISBN 978-0-87071-163-3 (paperback)
ISBN 978-0-87071-164-0 (ebook)

First published by Oregon State University Press in 2021
Printed in the United States of America

Originally published in 2003 by the University of Nebraska Press.

Oregon State University Press
121 The Valley Library
Corvallis OR 97331-4501
541-737-3166 • fax 541-737-3170
www.osupress.oregonstate.edu

Contents

Illustrations

Following page 102

Acknowledgments

A memoir is by nature subjective, so my first sources have been, of course, my own memories and those of my family. However, in researching the historical background for this book, I consulted as many written sources as I could find.

The industry and enthusiasm of Gil Davies and Florice Frank, founders of HiStory ink Books, have preserved many narratives of local and Forest Service history that would otherwise be unavailable to the public, and I urge anyone interested in the Klamath Siskiyou country to lose no opportunity to obtain any or all of their publications.

The Siskiyou County Historical Society, through its yearly publication, *The Siskiyou Pioneer,* has collected a large body of information about Siskiyou County. These are indispensable to any researcher. Some back copies are available from the Siskiyou County Museum in Yreka, California.

Many documents about the early history of the Klamath National Forest have been collected and preserved, thanks in large part to its irreplaceable archaeologist, Jim Rock. Federal laws such as the National Environmental Policy Act (NEPA) have led to the collection and publication of much data about watersheds and their history, which has proved invaluable in researching the history of places such as Beaver Creek. To Jim and all the other Klamath National Forest employees who have kept in touch over the years, a thousand thanks.

Hank Mostovoy, former Forest Service and Fruit Growers employee, provided me with information about Hilt's logging operations. A thousand thanks to The Mad Russian! Current photos of the Hilt townsite were taken by Neil Anderson, who made a long trip there on my behalf.

Portions of this book have been previously published as: (chapter 1, epilogue) "White Poplar," in *American Nature Writing 1998,* ed. John Murray (San Francisco: Sierra Club Books, 1998), 162–65; (chapter 1, epilogue) "White Poplar, Black Locust," in *American Nature Writing 1999,* ed. John Murray (Corvallis: Oregon State University Press, 1999), 184–96; (chapter 23) "A Long Death," in *American Nature Writing 2001,* ed. John Murray (Corvallis: Oregon State University Press, 2001), 190–201.

Permission to reprint photographs 2 and 3 has been granted by Neil I. Anderson.

Sections from *Memorable Forest Fires: 200 Stories by U.S. Forest Service Retirees,* ed. Gilbert W. Davies and Florice M. Frank (Hat Creek CA: HiStory ink Books, 1995) are reprinted with permission.

Part 1: White Poplar

CHAPTER 1

> I know an ash-tree
> Known as Yggdrasil,
> Tall tree and sacred
> Besprent with white clay,
> Thence come the dews
> That fall in the dales;
> It stands ever green.
> THE ELDER EDDA

My mother grew up in the tangled knot of forested mountains that spill over the California-Oregon border, hiding that artificial line, belonging to neither state. She was raised mostly in one small company-owned lumber town, a town born in a lost era when timber was cheap and the public domain lay open to those who could take it. But those days had been short, and even in the 1930s, Hilt, California, was a sort of relic, like McCloud and Gilchrist, Bray and Scotia. Some company towns, like nearby Klamathon, had vanished even before the Western lumber industry settled into the depression that began long before the Crash of 1929.

What saved Hilt when other lumber towns around it died were its ties to a larger, more stable industry: the growing and selling of citrus fruit. Hilt's owner, Fruit Growers Supply Company, was in turn owned by Sunkist, the huge southern California citrus growers' cooperative. Fruit Growers still exists and owns over three hundred thousand acres of timberland in northern California. It got into the lumber business before World War I to secure a reliable—and cheap—supply of wood for fruit boxes. Buying Hilt and its timber in 1910 was just a means to an end. Still, even before my grandparents moved to Hilt in the early 1930s, there were rumors that Fruit Growers was preparing to abandon Hilt. And Mother's parents raised her accordingly.

Get an education, they told her, as they sent her out the door of the unpainted company house and shooed her down the board sidewalk toward the school. Go to college, they told her, as she crossed the dirt road to catch the high school bus for the twenty-two-mile ride into Yreka, the county seat. Get a job in a place with more than one industry, they said. Grow up, get out, get away. And she did.

She was working in a bank in Ashland, Oregon, when she met a newspaper reporter named Dick Johnson. In 1947, she married him in Hilt's little white church, with all her childhood friends around her. I was born two years later in Boise, Idaho, where my father worked for *The Idaho Statesman.*

In the autumn of 1951, Mother put on her best tweed suit, left a note for her husband, and bundled my two-month-old sister and me into a taxi, then into an airplane that carried us over the Cascade Mountains to Medford, Oregon, where Mother's parents met us. We drove south over the Siskiyou Summit and down into the valley of Cottonwood Creek, and Hilt. Someone lifted me out of the car and set me down on the board sidewalk where my mother had learned to roller skate.

In later years I would come to believe that I remembered that day, remembered my first glimpse of Hilt. I thought I remembered Mount Shasta floating on the southern horizon, serene above the dark curved bulk of Black Mountain. I seemed to remember the long descent from Oregon and the way the forests of fir turned to oak and juniper on the round brown hills just outside of town. I thought I remembered the sight of the railroad tracks and the great mill complex on the west side of town, and the smoking tepee burners, and the smell of wood smoke in the cool fall air. I thought I remembered seeing for the first time the changeless circle of peaks that marked the boundaries of this new world: Mount Shasta, Black Mountain, Bailey Hill, Cottonwood Peak, Bullion Mountain, Shaft Rock, Hungry Creek Peak, Mount Ashland, Pilot Rock, Sheldon Rock, and Mount Shasta once more.

But all that came later. I don't, of course, really remember that day at all.

Our mother stood barely five feet tall and weighed a hundred pounds. Dark of hair and eye, with the straight, rather prominent nose of her father's family and the expressive mouth and firm jaw of her Danish mother, she was pretty but not beautiful; a neat, trim little person with a tough and practical mind and the ability to make that mind up into the consistency of reinforced concrete.

Our father had spent four years with this woman without learning this elemental fact about her. If coming home to find that note had not given him a hint of it, he was soon enlightened. He followed her to the airport, talked his way onto the passenger plane, and a heated conversation ensued that ended with Mother's firm refusal to disembark and a firmer suggestion from two large male passengers that Dick should take a hike. A few weeks after the three of us arrived in Hilt, Dick followed, with his older brother, our Uncle David. Although Mother actually liked David, her answer was still "no." The two men drove back to Boise, and Dick did what Mother wanted him to do. He sued her for divorce.

"He wants me to come back to him," she told our grandmother over yet another cup of coffee. "I can't. I can't respect him, and if I can't respect him, I can't live with him." Grandmother said nothing, only ran her tongue over her gold canine tooth, as she did when worried.

Old Mr. Ward, Hilt's town constable, came to the house one evening to serve the summons. After Mother had taken it from his hand and gone back to the kitchen to read it, he stood uneasily on the porch, looking through the screen door at Grandfather, his hat clutched in his hands.

"Let me know if there's anything I can do to help," Mr. Ward said.

"There's nothing," Grandfather said gruffly, and swung the big white door shut in his face.

"You shouldn't have been so sharp with him, Billy," Grandmother remonstrated gently, watching through the front window as the stooped figure made its way down the walk and out the wooden gate.

"None of his business," Grandfather retorted. "Now it'll be all over town tomorrow."

Since Mother had left her husband without alleging beatings, infidelity, failure to support, or habitual drunkenness, she was the deserter, the party at fault. To Mr. Tebbe, the lawyer her parents found for her, she would only state, primly, that "mental cruelty" was involved. In the end, she received sole custody of her children; our father got some vaguely worded visitation rights and the obligation to send Mother twenty-five dollars per month per child until we were eighteen years old. He was never late with a payment, but even in 1951, no one could feed and clothe and shelter a child on less than a dollar a day. Mother would have to find a job.

While she waited for her divorce to become final, Mother practiced her shorthand and typing. Then, armed with her bank experience, she marched down to the company offices and filled out a job application. Fruit Growers waited a decent interval—as though they actually had to give the matter some thought—before offering her a position as a clerk-typist. Her father was, after all, the foreman of one of the divisions of the lumber mill. He lived on Front Street, with the other bosses. His daughter and her children had nowhere else to go. It was better that she have a job.

By leaving her husband and moving back to Hilt, Mother had effectively ended her social life. Divorce in Hilt was a stigma, almost a scandal. Women who had known Mother for twenty years looked away as they passed her on the sidewalk and snatched their long New Look skirts away from her touch. Mother now lived on a knife edge, where a single breach of decorum or human mistake might plunge her into an abyss. So in the mornings, she was awake before the alarm. She dressed and disappeared down the sidewalk, her stocking seams straight, her size four high-heeled pumps polished, her brunette pageboy as flawless as her severely drawn red lipstick, and left us with Grandmother.

The Company executives who hired Mother did not, of course, tell her that they needed her as much as she needed them. In the expanding postwar economy, trained office workers were suddenly hard to find and harder to keep in a small and isolated place like Hilt. A young divorcee with ties to the community and children to support was, they thought, a godsend: she would stay, grateful for what the Company chose to give her. Fruit Growers, poised on the brink of the largest and most sustained lumber boom of the century, was optimistic about this new world. The hard times were over. War in Korea was sending the lumber market sky-high. Cargo ships, their holds full of wooden palettes, their decks crammed with dimensional lumber, followed the American forces overseas. And because no politician wanted to choke off the domestic housing market, demand rose still higher for the lumber left behind. Mother's job was safe, and all the old uncertainties were gone, but the Company took care not to tell her that.

The first geography that my sister and I knew was the house that Grandmother made into home for us, every day. The house sat midway down the long row of bosses' houses on Front Street. There were eight of them, larger than other Hilt houses, with bigger yards. My first real memories begin in that house, with Grandmother's routines: the yellow glow of kitchen lights shining through the half-open door of the back bedroom; the rustle of newspapers and the popping and crackling of kindling as she built fires in the two stoves; the sound of water running into sinks; the rattle of the percolator being thrown together, followed by the gentle swoosh of coffee perking, and then the warm smell of it.

From my small bed next to the wall in that bedroom, beside the big double bed that Mother now shared with Aunt Jo, I watched the morning, while my sister Elizabeth slept on in her crib. Adults stumbled past me on their way to the cavernous bathroom. On winter mornings, I put off going there as long as possible. The bathroom contained a claw-legged bathtub, a built-in towel cabinet, a sink, a toilet, and about a mile of the coldest linoleum in the world. For our winter baths, Grandmother preheated the room with a dangerous-looking portable electric heater that glowed red and smelled ominously of burning dust, but morning bathroom users were on their own.

At seven o'clock, our Aunt Jo, sixteen years old, scrambled into her coat and scarf, snatched up her books, and ran for the school bus stop at the end of the street. On the long porch that connected the Company store, the post office, and the Company offices, she waited with other teenagers for Howard Trivelpiece to guide the big yellow school bus out of its too-short garage and down the hill. At fifteen minutes to eight, Mother and Grandfather left for work. Grandfather put on his old gray sweat-stained work fedora before walking off across the railroad tracks to the box factory, half a mile away. He considered driving so short a distance a waste of gasoline. During the week, the old Chevy or its successor—the biggest Oldsmobile he could afford—was safely stored in a garage building several blocks away.

At eight o'clock, the mill whistles blew, and then the house was quiet. Grandmother allowed me to creep out into the warm kitchen and dress beside the kitchen trash burner. In winter, the insides of the twelve-paned windows behind the dining room table were covered with great fernlike traceries of frost, fading slowly under the heat of the front-room stove.

Grandmother sat at the big dining-room table, the breakfast dishes shoved into the middle, and wrote letters and shopping lists, or entered grocery slips into her big account book, and drank coffee. When Elizabeth began to fuss, Grandmother changed and fed her, and put her in her playpen in the front room, near the stove. Grandmother and I were the first to see Elizabeth stand, and crawl, and walk. To us she said her first words, and in the evenings, while Grandmother cooked, I crawled with her on the floor under the table and took dust bunnies away from her grasping hands and eager mouth.

Elizabeth and I came to know Grandmother's moods and stories and disciplines much better than Mother's. Grandmother comforted and praised us, punished and rewarded us, rocked us to sleep late at night when we were sick. When Elizabeth howled with the pain of her first teeth, Grandmother stayed up with her, ironing beside the couch, humming lullabies. When we misbehaved, Grandmother whipped us with tree switches in summer and a wet dishcloth in winter. Her punishments were swift and just. She fed us and talked to us and taught us prayers and songs. She was our parent. Mother heard our nightly prayers and sang nursery rhymes to us, sitting on the side of my bed. We watched her paint her nails or pluck her fine black eyebrows or curl her hair into little damp ringlets and stab them with crisscrossed hobby pins. In the evenings and on weekends we watched her iron blouses or sew on buttons. She read to us, so that I came to know my favorite books by heart and could recite them to Elizabeth or to obligingly admiring guests. The evenings were long in those days, and there was time for Mother and Aunt Jo to play canasta or trim each other's hair, time to pin up a skirt hem, time to talk and to read—all the time in the world.

We came to see our mother as an older, more sedate and cynical version of our flamboyant, talented, fascinating Aunt Jo, who banged out "Heart and Soul" and "Unchained Melody" on the old upright piano behind the door of our bedroom, or pushed back the front room furniture to practice her dance routines on the slick linoleum, while "Blueberry Hill" groaned out from the big Magnavox cabinet radio/phonograph. Mother and Aunt Jo were our friends, our sisters, our worldly older companions, and we pined to be like them: to wear brassieres and lipstick, to paint our nails. But Grandmother's magnificence seemed unattainable to us.

A bit shorter than Mother, but twenty pounds heavier, she had rather narrow hips, a wonderful deep bosom, and big hard biceps. Her pale skin was translucent and soft, her eyes a very light blue. She wore her light brown hair in a big coil on top of her head. When she took it down to braid it at night, it cascaded past her hips. When she washed it, it took two days to dry. When she pounded tough, round steak with a mallet in the hot kitchen, a sheen of sweat sprang out on her high forehead, her big arms vibrated, and we watched her with awe. In the beginning, our world was a woman.

She was born Martha Kristiane Vilhelmine Dittmar, a mouthful of a name that I used to recite to myself, enchanted by its rhythms. She came from Denmark to Iowa with her parents at the age of two and grew up in Clinton, a Mississippi River town full of immigrants—Germans, Swedes, Danes. She left school after the eighth grade and went to work in a Danish bakery, where on delivery days the men who drove the freight wagons unloaded hundred-pound sacks of flour onto her shoulders. She learned to mix and knead and bake pastries and to wait on customers. She worked ten hours a day for ten cents an hour.

Forty years later, as she pounded and chopped and kneaded food for us, the patter of the bakery counter seemed to come back to her, as she told us stories: of Jule Nisse, who left presents in the shoes of good children on Christmas Eve and received in turn a dish of clabbered milk sweetened with cinnamon and sugar, set out beside the hearth. She sang little songs in Danish to us, taught to her by her own mother Anna, an orphan who had been farmed out to a series of grudging relatives, then put to work as a goosegirl at the age of four. She told fewer stories about Niels, her father, a hot-tempered man who once horsewhipped a barber for shaving off his prized handlebar mustache as he slumbered in the barber's chair.

In Iowa, Niels was a carpenter and bricklayer in the summer. In winter he worked for an ice company, cutting great blocks of ice from the Mississippi, loading them onto sledges, and stacking them between layers of sawdust in insulated sheds. Anna raised chickens and geese, sold goose down and eggs and dressed birds, and had two more children—Sophie and Andrew.

Our grandmother Martha grew up in a world where safety razors and sanitary pads were miracles. Books were expensive and cherished. As she began the work of raising us, the flood of postwar consumer goods, rising higher and higher year by year, worried her. Everything had a price, she believed, and must be paid for in the end.

Sometimes, as she built the morning fires with wads of newspaper and kindling and slabs of rough firewood, she mused about her mother's life.

"The people Mama lived with," she said once, "used to send her into the woods with a big basket, to pick up branches and bring them home. We waste so much wood in this country, but sometimes it took her a long time to fill up the basket. All the poor people went into the woods in those days, picking up everything that fell, even little twigs. Wood is scarce there. Someday we'll have to do that, too."

Grandfather came into the room and laughed, waving his arm as if to encompass all the vast forest outside, all around us. "Right, Martha, right," he said, drawling out the words sarcastically. Grandmother stared at him for a moment, then turned away, struck a wooden match on the rough surface of the stove door, and lit the fire.

In our beginnings, the outside world came to us by newspaper and radio, but most palpably by the crank telephone on the kitchen wall. A local system connected Hilt's hotel, boarding house, hospital, store, Company office, and the Warren's saloon across the railroad tracks with the sawmill, box factory, railroad depot, and about fifteen houses in the center of town. Taped to the telephone's oak veneer box was a yellowing list of the combinations of short and long rings by which other telephones on the system could be signaled. Grandmother listened to the rings to make sure the call was for us (two short, three long) before picking up the cone-shaped black celluloid receiver and shouting into the tube mounted on the box. Of course, anyone else on the system could pick up the receiver and listen, too. Grandmother regarded this as ill-bred.

On that telephone, Grandmother talked to people named Emma and Mabel and Peggy. Elizabeth and I crowded around her as she stood on a short step stool and spoke about meetings and house fires and illnesses. One day she rushed into the house and snatched up the receiver, yelled at someone to get off the line, and began whirling the crank. A few moments later, long blasts of mill whistles signaled a fire, the short and long punctuations signaling its exact location in town. We ran outside and saw the big fire truck roll out of its garage at the end of the street and disappear around the corner.

Long distance calls were rare and were made from a telephone booth on the porch of the Company store. Of these calls I remember cold winter nights when we all trooped down the sidewalk together, our boots crunching on the snow as we walked the length of Front Street. While Grandfather bellowed at the operator, Mother and Aunt Jo linked arms and improvised a chorus line of stomps and kicks, their breath shooting white from under their scarved heads. Elizabeth and I bounced up and down the steps of the store porch, almost immobilized in scarves, coats, boots, and mittens. During the ensuing short, shouted conversation, everyone crowded close to the booth and yelled greetings to relatives far away.

As we walked home, our footsteps sounding hollow on the board sidewalk, I felt part of some great and rare and wonderful event. The stars winked, cold and happy, and Aunt Jo shone her flashlight up at them, and told me that the beam would go on, forever, to other worlds.

When I was six, men came to the house to pry the crank telephone off the wall, and every house in Hilt was hooked up to Siskiyou Telephone. A heavy, black celluloid rotary-dial telephone appeared on a small table behind the couch. All the lines were private, now, and rang only in one house, and no one could listen in on what Peggy said to Mabel. But by then there were also televisions, and game shows and soap operas, and it may have been that there was simply less to talk about.

By the time we were old enough to wander the yard unattended, the landscape around Hilt was the only one we remembered. Just beyond Hilt's houses and streets, heavy clay soils held the roots of oak and juniper and sagebrush. Between them, grass greened up after the fall rains, and flowers flourished in the spring. To the east, where the land grew dryer, the big mule deer of the high deserts bedded down beneath lava rimrock; to the west lived the small Pacific blacktail deer, creatures of the forest. In Hilt's valley the two met, drinking from calcium-laden waters that built heavy antlers on the bucks. They lived fat in thickets of poison oak and rested from the years of market and subsistence hunting that had wiped out their distant cousins, the Roosevelt elk. At night they stepped cautiously down the alleys, leaving the double crescents of their tracks behind.

Inside the peaks and hills were our neatly fenced yards, with the two-by-four tops of the fences just a little wider than the soles of our sneakers. Bordered with iris and Shasta daisies and tulips, the yards were little kingdoms of shade and green.

In the middle of each front yard, a board sidewalk ran down from the bottom of the porch steps to a wooden gate and out to join the wider boardwalk beside the dirt street. Our own front porch was draped with Virginia creeper, cool and breezy in the summer. Through the leaves and tendrils, hidden from all eyes, I peered at the yard next door, where the Rugers lived, and out at the hot street. An old couch and a huge old wooden trunk sat far back on the porch, under a bedroom window. In the trunk were neat stacks of woolen blankets and coats and sweaters, layers in a mothball pie. Sometimes we lifted up the lid and inhaled the luxurious smell of naphtha before climbing to sit cross-legged on top of the trunk, while robins and English sparrows sang in the branches of the enormous white poplar tree that dominated the front yard.

Four feet across at the base, with craggy, wrinkled gray bark, the tree had lower branches as thick as a pony's barrel and almost as low to the ground. We could dig our toes into crevices in the bark and scramble astride. Above our heads, the corrugated skin of the old tree smoothed out; the younger branches were mottled white. The leaves, bright green on the upper side, fuzzy white beneath, shook and spoke in the wind that spilled down Bear Canyon in the hot afternoons.

The old tree cast deep, dark shade, under which hundreds of sprouts sprang from the hidden, knobby ankles. White poplar, like its cousin the quaking aspen, likes to live in groves, with its clones. My grandfather mowed them all down, unsympathetic to such collectivist ambitions, but they only grew the faster for it.

In spring, long tendrils of catkins, oozing white cotton, drooped from the tree, falling in a litter, blowing into fence corners. In autumn, the tree dropped layers of leaves, six inches deep, until Aunt Jo led us, armed with rakes, in an assault. We scraped the leaves into huge crackling piles, shoveled them onto tarpaulins, and dragged them across the road to a strip of weeds beside the railroad tracks, where they were piled and burnt to gray ash under Grandfather's vigilant eye.

We called our tree a cottonwood, ignorant of its proper name. There were many of them in Hilt. Squat, weedy, wide branching, virtually unkillable, they mark the dry lands of the West, in old towns and homesteads from Montana to California, surviving and sprouting through heat and drought and shattering cold.

Utterly useless for fence posts or lumber, barely adequate as firewood, white poplars are solely and staunchly shade trees. I like to think that women spread them, giving little sprouts in cans of mud to friends and relations passing by. Genetic studies of white poplars might trace many forgotten family connections, of lineages scattered and lost in the vastness. Perhaps they are all related, segments of one great clone clan, like the people they sheltered.

Our own tree's base was surrounded by a ring of pale watersmoothed rocks, filled with soil intended to support flowers. The rocks came from a bar on the Klamath River, ten miles away. No flowers grew in the shady bed, however, and it would have been remarkable if any had survived the dimness and the alternate digging and stomping of playing children. My sister and I hid behind the tree trunk, or lay on its branches, ants crawling under our shirts as we watched the sky behind the rippling leaves.

To me, the tree was a live presence, aware of us, but dreaming always of the past, seeing and hearing long-lost memories in the air and the sunsets; memories gone but still a part of the valley, as much as the nighthawks wheeling in the summer dusk, as much as the old, old hills—but still just out of reach, just behind the blue mountains.

On the railroad tracks that ran through our valley, trains came rattling through, several a day and several in the night. On dark winter nights, they shook the windows, their comforting whistles dying away as the fire in the big front-room stove died down and the cooling house creaked, settling onto its uneven foundations. I listened to the breathing of my mother and sister and aunt, heard the coyotes talking to the retreating trains, waited for Grandfather's next vibrating snore, and knew that everything was as it should be.

Our father waited three years before he came to visit us, bringing a new wife. I sat again on his knee, five years old, showing off my green velvet dress, my toys, my accomplishments. On his other knee was Elizabeth, almost three, plump and pretty, with shining blonde hair in ringlets—laboriously achieved by Grandmother and a pair of curling tongs heated in the chimney of a kerosene lamp. Her enormous blue eyes gazed up in wonder at this stranger and proceeded to enchant him. With the sure instinct that even then told her how to work a room, she smoothed down the skirt of her pink and white dress over the crinolines beneath with one small hand, and jammed the thumb of the other into her rosebud mouth. Daddy melted like Jell-O nailed to a wall. I looked up from my flawless recitation of a Little Golden Book and realized I had lost my audience. I looked around.

On the couch across the front room from us a tall, black-haired woman with a small waist, a large black belt, and a well-filled sweater sat erect, making polite conversation with Aunt Jo, whose equally well-filled sweater seemed to be sizing up the stranger's. Pauline had married Daddy barely a month before. This was probably not what she had in mind when he told her he was a family man.

Daddy wore a suit and tie with the air of one who wears such clothes every day. Grandfather had, in deference to the occasion, put on his good blue suit, but he looked uncomfortable in it. I was fascinated by Daddy's hands—pale, like his face, like his yellow hair; slender, with long fingers; soft, without calluses. They were nothing like Grandfather's blunt, brown, callused paws. Grandfather's hands were clean and his nails were smooth, but they were the blunt, hardened hands of a lifetime of factory work. Wiry black hairs sprouted from the backs of his fingers, and the fingers were stained with nicotine.

Daddy's hands seemed never to have chopped wood or shoveled dirt or pounded a round steak. What they did instead, I could not imagine.

We sat down to a meal none of us ever remembered. As the afternoon's visit ended, we walked with Daddy and Pauline down the sidewalk and out into the street. A show-off to the end, I climbed up on the fence and walked along the top board. But it was Elizabeth that he picked up and hugged before getting into his car; bouncing her in his arms for so long that Pauline gave up waiting for him and handed herself into the passenger side door. As the car pulled slowly down the street before swinging around in front of the fire station and crawling past us again, Daddy waved, his pale eyes shiny with unfallen tears. Pauline stared straight ahead, looking fixedly at the road out of town.

Mother watched the car turn the corner by Alphonse's house with a look of contempt on her face. She blew out her cheeks, told me to get off the fence, and went inside with Grandmother and Aunt Jo. Grandfather, walking carefully on the damp lawn in his polished black "town" shoes, moved the lawn sprinkler. I rode my tricycle up and down the front walk, while Elizabeth rode behind me, standing on the rear axle. Grandmother came back out onto the porch and ordered us inside to change.

By the time we had scrambled into our summer play clothes—shorts and halters—Mother was standing by the big front-room stove, barefoot, in a sleeveless blouse and pedal pushers, having broken all records getting out of her high heels, nylons, and girdle. She stood drinking coffee and talking to Aunt Jo, who was similarly attired. It was ninety degrees outside and there had been no fires in the house since April, but the stove was a place of council, where Things Were Discussed.

"Dick just wanted to show off his new wife," Mother said, her lips hard and tight as she set her coffee cup down on top of the stove. A lit cigarette between her fingers mingled its smoke with the steam from the coffee, which gave off a pleasant, fruity bite. I recognized the aroma of the Christian Brothers brandy that Grandmother kept in the pantry. Aunt Jo, leaning against the white wainscoting behind the stovepipe, was puffing rather self-consciously on a Kool. Grandfather didn't like her to smoke, but found it hard to forbid it to her when Mother lit up, too. So Jo only smoked in the house when Mother did, and Grandfather pretended that he didn't see it. Mother usually smoked only after dinner, sitting relaxed at the table with the empty plates pushed back. She was not relaxed now.

Grandmother came in from the kitchen, a big white apron tied over her silky midnight blue dress. She held a mug of coffee in her hands. The ring of women closed around the stove, protecting their own, shutting Daddy and his world firmly out.

I never got over Hilt. It is as real to me now, when it no longer exists, as it was when I was three years old, or six, or twelve. I see it, sometimes, with an aching intensity that will not go away, so that the little valley beside Cottonwood Creek comes back to me in dreams and memories, and lives, all of it: mountains, and streets, and eighty-five brown houses, and a lumber mill, shimmering in the summer heat across the railroad tracks. I tell myself that Hilt was just another town, just a small part of a Western industry that ate its seed corn and died. This changes nothing. Hilt is still my home, though I lived there only eleven years, though I left it almost forty years ago, though it has been gone from the map since 1974.

The valley remains, though the streets are vanished under grass, lost except to eyes that still remember where they ran. The valley remains, though the houses that covered the low ridges between Watertank Hill and Cottonwood Creek are gone. I remember the length of every unpaved street, the width of every board sidewalk, the colors of dirt in every alley. I feel the clinging gumbo soil of the open field beyond Adobe Street beneath my feet and smell the scent of wet juniper on the little hills beyond the ballpark, where yellow violets bloomed in the brief wet springs. I remember how beautiful it was, this little world where every day we looked up at mountains covered with second-growth pine and fir, coming back from the fires and logging of the early twentieth century. In my childhood, laden logging trucks still groaned down the long grade into the mill, bringing logs from somewhere beyond our skyline. The tepee burners glowed by night and smoked by day, and we smelled the sharp tang of bark burning, and in the mornings the little Shay engine pulled flatcars of lumber out from the mill and onto the Southern Pacific tracks. And we thought it would never end. We were children, after all.

CHAPTER 2

I know where stands a hall
Brighter than sunlight
Better than gold . . .
Hosts of the righteous
Shall it inherit
Live in delight
Everlastingly.
THE ELDER EDDA

As each winter left us, a day came when Grandmother, as she began the laundry, flung open the back door, so that only the screen door stood between us and the brand new outdoors. I pressed my nose against the mesh and inhaled the sharp metallic smell of screen tinged with wood smoke. And one day, suddenly, the cherry tree in the back yard was in bloom, and the grass was greening up, and the pit of sand beneath the clothesline was warm and dry.

On laundry days, Grandmother filled up the wringer washing machine with hot water from a short hose attached to the faucets that emerged from the wall of the enclosed back porch. When the machine was full, she shut off the water, dumped in a good pour of Fels Naphtha soap flakes, and pulled out the big knob on the side of the machine. The agitator started turning with a clunk and a swoosh, and the sloshing churned the soapy water into a foamy froth that filled the whole house with its sharp oily scent. Grandmother picked up towels and sheets from a big pile on the floor and fed them one by one into the machine's open maw. When it looked about full, she grabbed a fat stick of kindling, worn smooth from years of duty, and jabbed it into the soapy mass, shoving her victims below the waterline. She dropped the Maytag's hinged lid, then stepped back into the kitchen to start on the breakfast dishes.

Dishes in the drainer, she returned to push in the agitator knob and lift the hot wet clothes out with the stick and shove them one by one through the wringer, which perched above the big stationary tub, now full of cold water. The soapy clothes, their suds squeezed out by the wringer, hissed into the tub and floated gently, expanding. Grandmother shoved them down into the cold water with her strong forearms, set the machine to going again, and dropped in more clothes. By starting with sheets and towels and ending with grimy overalls, she could do an entire wash using only the original tubful of water in the machine.

By noon, most of the laundry was out on the clotheslines, and lunch was on the table. Grandfather always came home for lunch. He worked his way quickly and noisily through sandwiches or a bowl of soup or leftovers, talking with his mouth full, and once every few days choking on inhaled food, sending Grandmother running to fetch water or slap him on the back, while I waited hopefully for him to turn blue the way Elizabeth did when she held her breath too long during a screaming fit.

In the afternoon Grandmother took us with her when she walked down to the post office to fetch the mail or to shop at the Company store in the same building. Long before we were old enough to roam the neighborhood and discover other children, Elizabeth and I were on speaking terms with an array of more or less elderly adults.

At the post office, we talked to the postmistress and her husband. Their name was Baumgartner, but they were always referred to as Bummy and Mrs. Bummy. He was a thin, bent man who sorted mail or bumped boxes around behind the wall of mailboxes in the tiny lobby while she waited on customers. He had white hair and wire-rimmed spectacles and there seemed to be something wrong with his knees. Mrs. Bummy was her husband in drag.

Young adults were not abroad much in Hilt in the daytime. Housewives with children in school shopped in the mornings. All the men were at the mill, except for the odd Company carpenter or plumber, or the butcher behind the meat counter at the store. He let us roam the cavernous storage area behind him, where cats sat on bags of oats and chicken feed, nursed kittens in boxes full of sawdust, and waited for scraps thrown from the door of the walk-in cooler.

Sometimes in the afternoons, Grandmother took us to visit Mrs. Sife, who lived near the top of one of three long rows of houses that trailed down from the base of Watertank Hill. She let us wander around her front room and stare at her collection of elephant figurines poised on a series of tiny shelves nailed to the flowered wallpaper. She dusted the elephants much more regularly than the rest of the furniture. On the stained, flowered chintz sofa covers, small irritable dogs slept uneasily.

Mrs. Sife always entertained us in her kitchen; no one except formal company ever came to the front door. We entered the back gate from the alley, walked up three little steps, and into an overwhelming smell of cooked cabbage. Grandmother believed cabbage suitable only for coleslaw and hated the stench, which Elizabeth and I found merely exotic, but she sat at the kitchen table and drank a leisurely cup of coffee, while Mrs. Sife poured Elizabeth and me glasses of Kool-Aid, which was approximately the same color as her hennaed hair.

Mrs. Sife's back yard was a warren of small dirt paths between raised vegetable beds. Mr. Sife brought home wide boards from the mill to build them and the compost bins in which his wife deposited leaves, chicken manure, rabbit droppings, and kitchen waste. It must have been the only non-Italian biodynamic garden in Hilt, and it grew Swiss chard with leaves the size of palm fronds and parsley so lush and green that we picked off little pieces and nibbled them, the fragrance shooting up into our sinuses. She raised far more chard and parsley than she and Mr. Sife could eat, so she ordered Grandmother to take as much as she wanted, anytime. By the time I was seven, Grandmother knew she could get us out of her hair on a summer day by handing us a shopping bag and sending us up to Mrs. Sife's for parsley. We never met anybody else in her garden on these expeditions, nor did we ever see any other housewife besides our grandmother sitting in Mrs. Sife's kitchen drinking coffee. Not many people visited Ethel Sife. People said that before she married Mr. Sife and moved to Hilt, she had been a whore in Seattle.

Mabel Hall, the other neighbor on whom Grandmother paid afternoon calls, would never have gone to visit Ethel Sife herself, but she was happy to accept greens from the scarlet woman's garden, as long as Grandmother brought them. Mabel and her husband, Art, lived three doors south of us, and although Art had been a sailor, Mabel had certainly not met him in the way of business. Mabel still wore her hair bobbed and marcelled in the style of the nineteen-twenties. Tall, lean, and cool, with faded auburn hair, a large nose, and a harsh voice, she was our introduction to the formal etiquette of visiting. She was not fond of children, and in her house we sat quietly and did not wander about. As she talked with Grandmother, her fingers flew over a piece of tatting or crochet, rapid as a spider's legs wrapping a fly in silk. Every available surface in her house was covered with her needlework—doilies and antimacassars on tables and the backs and arms of chairs; a glimpse of small pink dolls through the open door of the bathroom, their crocheted hoop skirts spread modestly over rolls of toilet paper. Just in case she ran out of ideas, the magazine rack beside her chair overflowed with back issues of *The Workbasket.*

Art's large brown leather easy chair occupied one corner of the front room; it was flanked by a polished brass spittoon that was always dry and clean in the afternoons. The room was permeated by the smell of the fat brown cigars he smoked. He was the only man in town who still wore an old-fashioned cloth cap, by the 1950s a trademark of elderly golfers. An Alabama sharecropper's son, he had joined the navy to escape a life of chopping cotton. On a destroyer in the Far East, he rose to machinist's mate, and in the 1920s sent stacks of postcards home from China and Japan. Mabel eventually gave them all to Mother, who kept them in a large shoebox and sometimes let me look through them on rainy days: highly colored glimpses of a dreamy world where manicured gardens and strange blossoming trees grew against a backdrop of mountains like misty thimbles.

Art raised pigeons in a dovecote behind his woodshed, so that he could have squab; it was the only dish, Mother said, that Mabel could prepare really well. Certainly her crumbless, shining kitchen looked and smelled as though no one ever cooked there.

Art's politics consisted of a hatred of Yankees and "niggers," defined as anyone with skin darker than his. He found the sight of Hilt's one Native American family offensive and snorted about the Hawaiians who lived in the apartment house at the foot of Front Street, hut his sensibilities were spared the sight of African Americans, for the Company simply would not hire a black man. We saw them only at a distance, maintaining the railroad tracks or riding in the engines and cabooses of trains clacking slowly through town. We waved, and they waved back.

When Art's skills as a machinist drew an offer of a promotion if he would move to Weed, some fifty miles to the south, he refused. In Weed and nearby Mount Shasta City and McCloud, black people worked in the lumber mills and sent their children to the public schools and thought they were as good as anybody.

Sometimes a black man drove into Hilt and walked up the steps of the Company office and asked about work, and from her desk Mother would see Bill Tallis come out of his office in the back and tell him that Fruit Growers wasn't hiring now, but that the mill in Weed might have some openings. The man would nod and say thank you and walk down the steps and drive away, understanding the code inside the familiar words.

In summer, the long hours between the afternoon errands with Grandmother and supper stretched nearly forever. As red-tailed hawks and turkey buzzards soared high above the town in the heat, the afternoon breezes gradually strengthened until, at about three o'clock, a wind came out of Bear Canyon, below Mount Ashland, with a sudden roar. Dust devils rose in the dry streets, clothes snapped on clotheslines, and Grandmother lay down for her nap. I stared out the bedroom window, sweating and bored, watching leaves and paper whirling aloft into the burning sky. In her crib, Elizabeth snuffied, damply unconscious.

At 4:30, the high school bus pulled up in front of the store with a sigh of brakes, and Aunt Jo and her friend Rosalie Graves came down the sidewalk. Sometimes they both came in, to fling their books and binders down, eat anything not nailed down or boiling, and play records. When the five o'clock mill whistles blew, Rosalie trotted home, and Jo began setting the table for supper.

In good weather, I climbed on my tricycle as the whistle blew and pedaled furiously down the sidewalk toward the Company offices, stopping at the end of the sidewalk, watching the homebound caravan of cars emerging from sawmill and box factory, from lumber yard and planing mill and loading dock. Mother came down the steps of the office across the street, and I pedaled home with her. By the time we reached our gate, I could see Grandfather walking toward us, crossing the railroad tracks. We ran up the steps and into a house smelling of home and food.

Grandfather washed up and immediately sat down with a can of beer and the *Siskiyou Daily News.* We ate supper within half an hour, and afterward Grandfather sat smoking and reading the paper and listening to the radio while Grandmother and Mother and Aunt Jo washed and dried the dishes. The evening passed in a circle of light and warmth and the small maintenance jobs of women: the mending of bra straps and the sewing on of buttons, the pressing of a dress or skirt, and always the women's talk that bound them together.

Sometimes, a couple of hours after supper on Friday evenings, Aunt Jo excavated the electric popcorn popper from a lower cupboard and made popcorn and cocoa for us while we listened to *The Lone Ranger* on the radio. Hunkered down on the floor, I stared at the eerie green glow behind the dial, imagining that if I stared hard enough, images of the tale would appear in the interior light. Surely if the characters were in there talking, there had to be some way to see them.

When I was five years old, we got a television set, and after that, Friday nights meant "the fights"—boxing matches, which Grandfather turned up loud. They provided a background noise, which, rather than lessening the women's talk, made it easier. What Grandfather couldn't hear, he would never know.

I think that was when I first learned that people could live in the same house and yet inhabit different worlds. Elizabeth and I knew, without being told, that the talk we heard in the kitchen was not for Grandfather's ears, just as we learned that the pinup calendars we saw behind the bar up at Gino's Stateline Liquor Store were not something we should talk about to Grandmother. And still, I wonder how we learn them, the many unspoken codes that we never afterward forget.

CHAPTER 3

Except for the War till the day he retired
He worked in a factory and never was fired.
W. H. AUDEN

Company towns, a fixture at the turn of the last century, have all but vanished from the West. These were not the coal towns of Appalachia, where men worked under conditions close to debt peonage, nor the dark and gritty steel towns of the Ohio Valley, nor even the mining towns of Colorado and Idaho and Montana, where a violence now forgotten marked the Progressive Era. Instead, Gilchrist, in central Oregon, and McCloud and Hilt and—now the sole survivor—Scotia, deep in the coastal redwoods, were new and clean and bright for a few years, until the unpainted houses mellowed to soft gray-brown in the sun and wind. These were places where people wanted to live and work, where companies wanted workers to stay. In the Pacific Northwest, a good millwright was worth keeping, and a top sawyer might quit and be on the next train if offered four cents more an hour fifty miles up the line. It was worth a company's while to offer a house, worth it to build a school, worth it to promise a pension to anyone who would stay for thirty years.

Until the 1950s, babies were born in Hilt in the Company hospital, delivered by Dr. Langer and later by Dr. Schlappi, the Company doctors. Until the very end, people bought milk and bread at the Company store and mailed their letters in the post office next door. Company carpenters fixed broken porch steps on the houses that workers rented for twenty dollars a month. On Monday mornings, Orson Coleman, the Company garbage man, hauled away your trash in a Company dump truck. Company plumbers crawled under your house every spring to turn on the outside water valve, nestled in a cedar box filled with sawdust against the winter cold. Company trucks delivered loads of mill ends and slabs for winter fuel to your back door for a nominal fee. Cars were parked in Company garages and filled up with gasoline at the service station run by Bob Trinca, who leased the land from the Company. We talked about Our House and Our Yard, even as we knew that none of it really belonged to us, for Hilt was owned, down to the last splintering board on the sidewalks, by the Fruit Growers Supply Company.

Fruit Growers could bring you into the world and deliver wood to warm you and give your parents work to feed you. The only thing it would not do was bury you. It drew the line at that, for in a town where the Company owned everything and where everyone worked for the Company, the dead had no place. A graveyard might have given the next of kin some claim on the land that they could never own in life. So although eulogies were spoken at the little white community church, Hilt people were buried in Hornbrook, eight miles to the south, or Ashland over in Oregon, or in Yreka.

Sometimes the Company allowed retired mill workers who had nowhere else to go to live on in Hilt, but that was different; that was charity for a few whose children worked in Hilt. Old Man Marin and his wife, for instance, lived in a tiny house, built by their sons, behind the Company store. Another old couple remained in the house on Adobe Street where they had lived for forty years. But the Company had no fear that Hilt would become a retirement village. After World War II, with housing loans easy to get, especially for veterans, new houses began climbing the hills above Yreka and Ashland, and ever-younger workers thought about buying one of those houses and commuting to their jobs.

If the twenty dollars rent to the Company reminded every family that their house was not their own, the reminder was gentle, and children knew nothing of it. Hilt was ours, and we thought not at all of the cold fact that Hilt existed for the sake of the great mill complex across the railroad tracks—a world of tiered roofs enclosing floors covered with screaming machinery—sawmill, planing mill, box factory, and dry kilns. Outside were the mill pond and log decks and lumberyard. Without the mill and the logs, snaked out of the pond one by one and fed into the saws, there was no work, no town, no community, no *us*. Everyone—whether they worked in "the plant" or not—lived by the rhythm of the mill whistles that called the men to work, signaled coffee breaks and lunch, and sent them home at five o'clock. The whistles ran on steam created by the heat from burning waste wood and sawdust—steam that made some of the electricity that ran the plant.

Our town's namesake was long dead: John Hilt of Illinois, who came to California in search of gold. In 1852, he and a partner, who called himself Smith, walked four hundred miles north from the played-out mines of the Sierra Nevada to Yreka, largest of the far northern gold camps. Twelve miles further on, they crossed the Klamath River at Henley, near the mouth of Cottonwood Creek.

Hilt and Smith made $150,000 mining gold on the Klamath. At Henley, John courted and married a girl from Michigan. Hilt and Smith were prudent; they bought a water-powered sawmill on the West Fork of Cottonwood Creek and sold lumber to other miners. In 1877, Hilt bought out his partner, moved the sawmill two miles up the creek, installed a newfangled circular saw, and converted the operation to steam power. For almost a hundred years, there would be a lumber mill in the valley of Cottonwood Creek.

Cottonwood Creek was mostly a route to somewhere else, but a few people stayed. A mile north of Grandmother's clothesline was the site of Cole's Station, where stagecoaches once changed teams before making the long, steep climb into Oregon. A family named Cavin lived on the West Fork of Cottonwood Creek even before John Hilt arrived in California. Silas Shattuck raised beef for the miners and achieved a brief local notoriety when he found some mail sacks discarded by Black Bart, the famously nonviolent stagecoach bandit, after one of his two holdups in the area.

Miners built a road from Cottonwood Creek over the ridges into Beaver Creek; from it, trails branched off to the mines. After the railroad came in 1887, the old stagecoach route past Cole's fell into disrepair, but early in the new century, a saloon keeper named Slim

Warrens loaned Siskiyou County some money to build a new road. Then he built a saloon called The Diamonds near the railroad tracks. Oregon was dry then, and the train and the new county road brought customers to his saloon. He ordered an enormous cherrywood bar from the East, and had it shipped around the Horn to San Francisco. Its brass foot rail and spittoons and acres of back mirrors and lights were set off by wrought iron tables and chairs and pool and billiard tables.

In the 1940s and early 1950s, men still walked into Warrens's toward that bar, passing a barber shop and a magazine rack full of *True* and *Male* and *Police Gazette.* Fedoras tilted to the back of their heads, they rested a foot on the brass foot rail, drinking beer in the cool dimness under the high ceiling and the lazily turning fans. But they were not the footloose hell-raising loggers and railroaders of the old days, who rode the trains down to Warrens's on Friday nights and caroused until Sunday morning. They drank their beer and then trudged out the door and across the tracks toward home, to their wives and their children and their company houses.

John Hilt sold lumber and shakes to miners until the 1890s, prospering enough to encourage two of his nephews to move out from Illinois and settle near him. By then he and his wife were old, and just as he put his mill up for sale, the country slid into a long depression, the Panic of 1893. Not until 1902 would four Oregonians buy the business and call themselves the Hilt Sugar Pine Company.

The new owners, as optimistic as the century, bought new equipment that turned out 35,000 board feet of lumber a day. (A board foot is equal to a piece of wood one foot wide, one foot long, and one inch thick.) They loaded their sawn lumber onto wagons and hauled it to a lumberyard near the Southern Pacific depot, a mile south of Cole's Station. They built a small planing mill to smooth the rough lumber into finished boards, then built a store, an office, a cookhouse, and a house for the yard foreman. They nailed up a sign beside the railroad tracks, bearing the new town's name.

The partners ran out of money just in time for the Panic of 1907 and sold out to the Northern California Lumber Company. NorCal built the box factory, which still stood fifty years later, and new railroad tracks out to John Hilt's old mill on Cottonwood Creek, and three dry kilns for curing green lumber. NorCal also expanded the town, building a water system to bring water four and a half miles down the West Fork of Cottonwood Creek. They dug sewer lines and built Hilt's first hotel and three streets of houses, which from a distance looked like Monopoly pieces thrown down on the treeless brown slopes. From Southern Pacific, which had been granted every other section of land for over twenty miles on either side of the railroad route (a subsidy from the federal government, to reimburse them for building the railroad), NorCal bought land for the town and thousands of acres of timber. The gold mines, primary buyers of lumber, were slowly dying, but a whole new market for wood was emerging in California.

Far to the south, as great irrigation projects watered the Sacramento and San Joaquin Valleys, orderly groves of fruit trees began to grow in the dry summer heat: peaches and plums, oranges and lemons and grapefruit. The fruit was shipped all over the nation in refrigerated rail cars, the oranges and lemons nestled in boxes made of thin slats of pale pine, cut from the California forests that had been given away to Southern Pacific and other railroad companies.

The Northern California Lumber Company built the first logging railroad lines into the woods west of Hilt. But railroads were expensive, so they borrowed $100,000 from Fruit Growers of Southern California, a cooperative of citrus farmers, hoping for a steady market for their box shook. The loan built the railroads, but by the time the lines were complete, NorCal was bankrupt. In 1910, Fruit Growers took over the logging operations and the sawmill. They would stay in the lumber business for the next sixty-three years.

In time, Fruit Growers of Southern California became Sunkist, and their supply subsidiary was called Fruit Growers Supply Company. The Company owned all the land and buildings in Hilt except the Southern Pacific depot and Slim Warrens's saloon. They built a new sawmill and lumberyard close to the box factory in 1911. By 1913, they were operating the entire plant ten hours a day, six days a week.

Out in the woods, Fruit Growers extended the logging railroads into Hungry Creek, Grouse Creek, and up to the heads of Nicklewaite and Cottonwood Creeks. The narrow gauge lines snaked out to Long John Creek and Red Mountain Creek and to the very sources of Beaver Creek. The logging camps worked six nine-hour days a week. Eventually, there would be twenty-two logging camps. The last one built was on Long John Creek; by 1930 there were forty miles of railroad grades in the Beaver Creek drainage alone, and twenty trestles soaring over creeks and draws.

The Company installed a newer, larger steam engine to run the sawmill, connected to an electric generator that produced enough electricity for the mill, with some left over to light the town two days a week, in those days before the California-Oregon Power Company extended its lines to Hilt.

The town hummed with youth and optimism and naive attempts at glamour. In 1920 the Sunkist Hotel went up on the corner where Front Street met the county road that looped down from the state highway. The Grand Opening in June was the social event of the year, some said the decade. Ladies came in their new shorter skirts; men wore their best suits and most uncomfortable high collars. In the dining room, long tables held food for a buffet supper, and the celebrants admired the huge stone fireplace and the lounge's comfortable Mission-style furniture. The dancing went on until dawn.

Although electric lights undoubtedly prevented some fires, everyone still cooked on wood-burning stoves, and the uninsulated chimney flues frequently caught fire. In 1924, a fire in the kitchen of the apartment house behind the Sunkist Hotel spread swiftly across the alley, and in an hour the apartments, the hotel, and ten other houses were gone. The Company store across the road was only saved when Tuffy Elmore climbed onto the roof and slapped out spot fires with wet burlap sacks.

When the smoke cleared, the townspeople saw that the roses in the hotel yard were unharmed. They dug them up and moved them to their own yards, so that in 1932, when my grandparents moved to Hilt, some of the shrubs stood ready to greet them, in front of the shabby little unpainted house just behind the old hotel site, blooming in a riot of pink all over the front fence.

In 1974, a meeting of the Siskiyou County Historical Society called for suggestions on the subject of next year's yearbook. Frank Graves practically broke his neck getting up to the podium before anyone else. He had been researching the town's history since his retirement from

Hilt in 1961, gravitating to the public library and the county museum with an unexpected hunger for preserving the story of the town. Now he set to work, possessed, for the only time in his life, of a guaranteed audience. The historical society's yearbooks—each dealing with a single aspect of Siskiyou County history—always sold out their entire printing runs.

Frank was, for forty-one years, successively tally man, grader, and finally yard foreman for the Company. His life revolved around the stacks of graded lumber that stood in neat rows beside the railroad tracks, waiting to be loaded onto the rail cars, or used for repair or construction in the town, or sold to the box factory. Each division of the plant kept its own books and made its own profit—or loss.

Frank was our neighbor. He lived three houses north of us on Front Street with his wife and three daughters. His marriage to Mildred brought him entry into what passed for gentry in Hilt, for she was a great-niece of old John Hilt's wife, Nancy, while Frank was just a boy from central California who had come to Hilt looking for work. Mildred's family hadn't thought Frank good enough for her; an opinion which, it was generally believed, had come to be shared by Mildred herself.

So, in his old age, Frank began to write. He wrote about the dump truck that Fruit Growers bought in 1924. He wrote about the number of men who worked on the sawmill carriage: one setter and two doggers. He wrote about the high tram that delivered lumber to the box factory from the sawmill, on tram cars drawn first by horses and later by tractors. He wrote about how horses and tractors sometimes fell off the tram docks, maiming men and animals alike.

He wrote of the days of logging with steam donkeys, when all the logs were hauled to the mill by railroad engines over a labyrinth of railroad grades snaking into the woods. He wrote of the crews of Swedes who built and maintained the tracks, and of the first gas shovel, and of the conversion of the box factory to electricity in 1916. He wrote about the four dry kilns that aged the green lumber. He remembered for his readers the sawmill as it was in 1920, with its nine-foot headband rig and the big gang saw, replaced in 1921 with a pony-band head rig.

Frank's memories flow on through the yearbook, happily; how the mill was operated from a line shaft driven by a great steam engine; how the gang saw had its own engine to pull it; how four small logs could be cut at one time by putting two logs on top of two others; how the pony band allowed better grades of lumber to be cut from the same logs.

When Frank first arrived in Hilt in 1920, the big donkey engines steam powered machines that moved logs from stump to loader—still ruled the woods. Mounted on wooden sleds, with a water tank and a wood rack, they were pulled from one logging show to another all summer. In winter, the wooden sleds were replaced, built anew from split logs. The donkeys and the railroads would disappear beginning in the 1930s, as caterpillar tractors and logging trucks filled a new niche opened by the first logging roads.

All this, and much more, Frank tells. And under his editorship, other voices add more details about Hilt's people: about the resident manager's wife who opened Hilt's first school; about the dances and the movies on Saturday nights; about the unpopular foreman who was wrongly accused of bootlegging. But one person is absent. Of Bill Roush, my grandfather, foreman of the box factory, and neighbor for thirty years, there is nothing save one bare, unavoidable mention in a list of plant foremen.

They looked rather alike from a distance: two graying, bald men in sober Eisenhower jackets and stained work fedoras, walking to work every morning and back again every evening. But each preferred to pretend the other did not exist, although for many years they both attended the arcane rituals of the same Masonic lodge. Bill Roush was foreman of the Company's box factory for over twenty-five years, his office separated by an inch or so of hoard from the shrieking planer, which gradually rendered him deaf in the upper registers. The box-factory office was small and dim and cool, with a big roll-top desk and a wooden swivel chair with a cracked, black leather seat. The walls displayed a panorama of pinup calendars, on which some thirty years' worth of rather modestly posed women, wearing little but large hats, were displayed. The office was suffused with the smells of sawdust, lubricating oil, and tobacco smoke.

The ideal wood for building light but sturdy orange crates is sugar pine. Old-growth sugar pine is straight-grained, light in weight and color, and splits easily. Sugar pine is only one species, and not the dominant one, in the vast mixed conifer forests of northern California.

Scattered among Douglas-firs, ponderosa pines, incense cedars, and true firs, it reaches heights of over two hundred feet, rising on a trunk as large as seven or eight feet in diameter. As it grows, it sheds its lower branches—a phenomenon known to foresters as self-pruning—which gives the tree a straight, smooth bole unmarred by knotholes.

Today, old-growth sugar pine is almost unobtainable and is worth thousands of dollars per thousand board feet. But in the box factory at Hilt, the lovely, malleable, straight-grained wood was sawn into the small lengths called "box shook" and delivered in prepackaged bundles to the warehouses in southern California, where they were nailed up into fruit boxes as needed.

Each morning, as Grandfather walked into that office, he was confronted with the necessity of buying pine boards from the yard at so much per thousand board feet and turning them into an assigned number of those little bundles of box shook. And as he and Frank walked their parallel paths across the railroad tracks each morning, each believed, with absolute certainty, that the other man's chief joy in life was to cheat him. To Frank, Bill Roush was a *parvenu,* a newcomer who had slunk into town broke and hopeless in 1932 and had run roughshod over others to get a foreman's job. To Bill, Frank Graves was just a bum from the flatlands who, having married the next best thing to the boss' daughter, was jealous of men with more ability than he.

I can see them still today as they looked in those days when I first knew them, forever walking away to the mill, on mornings so beautiful that even my child's heart ached to see how perfectly the sun struck the sharp blue pointed peaks. I don't know how I could have been unaware of the hatred there, a hatred nurtured by the fact that there was just enough truth in each of their opinions to feed the hate and keep it alive, until it outlasted their jobs, their ambitions, Hilt itself, and finally their own lives.

When my generation looks back at the 1920s, it thinks of the Jazz Age and F. Scott and Zelda in Paris and bootleggers and buying stocks on margin. But in Hilt, as in much of rural America, the optimism of the early 1920s quickly turned sour, dousing the flickers of bright hope. Despite the post–World War I avalanche of automobiles and electric appliances, the Western mining industry never really recovered from the Armistice of 1918. Grain and livestock prices dropped and dropped again. Fruit Growers paid some of its mill workers twenty-five cents an hour, and they were glad to get it, glad to have a Company house to live in and a Company store where Company coupons bought flour and meat. Many people in Hilt raised big gardens and kept chickens. Sometimes, when orders for lumber or boxes ceased entirely, the sawmill, box factory, and woods operations simply shut down. Those who thought they could do better somewhere else moved on. But most people stayed, because as soon as the orange harvest started, the orders would begin again, and in the meantime you had a place to live. In Hilt, then, there was little to choose between 1928 and 1932, and for Bill Roush, there had never been a boom time, never a Jazz Age, at all. There had only been the Great War and its brief, happy aftermath, which had taken him out of the mill towns of his childhood and briefly shown him a larger world.

Bill Roush's father was an ineffectual little man who failed at farming and shopkeeping and was finally bullied by a desperate, stubborn wife into buying a small candy store in Rainier, Oregon. Nominally owned by her husband, the business was in fact run and its decisions made by Lena Roush. Permenias just worked there. In family photos, he looks thin and sheepish and slightly lost; she stands heavily on widespread, sensibly shod feet, dressed in black, her gray-streaked hair skinned back in a bun. She looks as wide and solid as a refrigerator. She outlived him.

Bill was their seventh child, born on a farm in eastern Nebraska in 1897. Even as the big healthy infant drew his first breath, the farm was failing, crushed by drought, grasshoppers, the Cross of Gold, and the fact that Permenias simply wasn't a very good farmer. He named the boy William Jennings Bryan Roush, hut not even this tribute to the Great Commoner could make Nebraska tolerable. The Roushes left soon after Bryan (as his family always called him) slipped past his first birthday. They fled to the cool forests of western Oregon, where Permenias left his family in a rented house near the village of Canyonville and promptly disappeared into the logging camps.

Lena, bred in Missouri, was alone in a strange country with seven children and one sack of white flour. She sent her oldest son off to look for work; Charlie found it in a coal mine in Washington. The other children were sent to school through the eighth grade—a frightening indulgence, considering the state of the family finances. But by the time Bryan was thirteen and out of school, his sister Fanny was married to a lumber buyer in Mount Shasta City, in northern California. His brother, Grover, five years his senior, hoarded with Fanny and her husband, Clare, and worked in a local lumber mill. He also hustled pool in saloons and had begun to court a respectable German Catholic girl, to the horror of both families. Bryan followed Grover to Mount Shasta—then known as Sisson and in the shadow of the great white volcano for which the town was named, he was soon working in a lumber mill and learning to smoke and drink under his brother's tutelage. But he valued security too much to gamble his wages on skill with cards or dice or a pool cue. He wrote dutifully to his mother once a week, sent penny postcards to his little sisters, Mae and Frances, and worked six days a week pulling greenchain at the mill.

By the time he was eighteen, he was five feet, eight inches tall; a compact young man with strong, well-shaped, calloused hands already decorated with black hair and nicotine stains. In the fashion of the day, he combed his hair straight back from his high, square forehead. His eyes were dark brown in a tanned face, his nose broad with flaring nostrils, and he sported a large mole on his upper lip.

He joined the army in 1917, took out a life insurance policy payable to his mother, and prepared to go overseas. But the war ended before he could ship out; he got no further than the southern California desert, where he rose to corporal and received a campaign medal inscribed "For Service on the Mexican Border." In the army, Bryan's friends called him Billy. His best friend was Blackie Sanchez, a cocky raven-haired Californian with roots deep in the old Spanish Empire—or so he claimed. Discharged and flush with army pay, Billy and Blackie jumped a freight train heading east. They had a destination, although they took their time getting there—Billy had a letter from Clare, addressed to the manager of the Curtis Sash and Door Company in Clare's hometown of Clinton, Iowa. The army buddies worked there for almost a year and on Saturday nights went to dances at the armory, dancing until dawn with German and Scandinavian immigrant girls.

Homesick at last for the West, the young men rode freight trains south to New Orleans, stopping to buy a ticket before passing through the chain gang counties. From Louisiana they rode across the Southwest toward the dreamland of Los Angeles and finally stepped out into the glare of southern California, with all their lives before them. In Billy's shirt pocket, carefully folded, was the name and address of a Danish girl, who had waved good-bye to him from the Clinton station, and told Billy he might write to her.

Fifty years later, Grandfather still talked about that glimpse of America from a train, the long journey back through a Midwest his family had abandoned, as the great adventure of his life. He was never to take a longer trip. Sometimes he told us about his first view of the Ohio River, mingling with the Mississippi at the toe of Illinois. He passed it on an October evening when the citizens of Cincinnati, far up the dark Ohio's waters, were celebrating the victory of their Reds over the Chicago White Sox in the 1919 World Series. The Reds' center fielder, Edd Roush, was a distant cousin. I didn't understand why this moment had cast itself so firmly in Grandfather's memory until I read Lawrence Ritter's book about baseball in the early twentieth century, *The Glory of Their Times.* As I turned the pages of Ritter's interview with Edd Roush, I came upon a photo of him, taken in 1916. Looking out at me was my grandfather's face.

Billy and Edd both traced their ancestry to a German named Johann Rausch, who fled Europe for America in the 1750s. In Pennsylvania he married a woman named Susannah, and by 1773 he was prosperous enough to take up a four-hundred-acre land grant on Mill Creek in Shenandoah County, Virginia. The Rausches—who swiftly Anglicized the name—seem to have been tanners by trade; at any rate, Johann's son Philip owned a tannery on a stream near Mount Clifton. Philip and his seven brothers all fought in the Revolutionary War, but even before that, one of them—Jacob—had traveled beyond the Appalachians.

His journey was connected to a small affair later known as Lord Dunmore' s War. When George Washington toured the Kanawha and Ohio River valleys in 1770, looking for land to give away to veterans of the French and Indian Wars, he liked the country. But when military surveyors arrived several years later, bloody skirmishes erupted as the actual owners—the Indian tribes of the area objected. Virginia's House of Burgesses authorized Lord Dunmore, the last colonial governor of Virginia, to raise a militia.

The force was divided into a northern division, commanded by Lord Dunmore himself, and a southern division, commanded by General Andrew Lewis. The two divisions planned to meet at the mouth of the Great Kanawha River, south of the Ohio, and advance together against the Indians, but the Indians crossed the Ohio River first and attacked Lewis's force near Point Pleasant, West Virginia. Seventy-five militiamen were killed and the whites were on the brink of defeat when the death of a warrior named Pucksinwah discouraged the Indians; they withdrew across the Ohio with only twenty-two killed and eighteen wounded. Neither side had much stomach for further encounters, so the adversaries met a few weeks later and agreed that the Indians would stay north of the Ohio River while the whites kept to the south. Lord Dunmare's army declared victory and went home. Jacob Roush told his family about the beautiful Ohio country, and they remembered. The promise of the Virginians to stay south of the Ohio River turned out to be worth what such white promises to Indians usually were. The later service of the Roush brothers and thousands of others in the Revolutionary War was rewarded with land grants in the Ohio valley, and by 1796 all of old Johann's children except Mary Magdalene, his only daughter, had floated down the Ohio River and settled on its north bank with their wives and children.

In these new lands beyond the Appalachians, the third generation of Roushes grew up and married women from other tribes: Pence and Frye and Smith and Scott. Within two generations the Roushes formed yet another strand in the amorphous herd of European settlers swiftly overwhelming the lands between the Appalachians and the Mississippi River. Seven generations later, our branch of the family had reached the Pacific Ocean, not out of Manifest Destiny, but out of hunger. Their backs against the ocean, they became part of a great army oflandless laborers in the logging woods, cutting down one of the last great temperate forests on earth for a dollar a day.

CHAPTER 4

For what do we live, but to make sport for our neighbors, and laugh at them in our turn?
JANE AUSTEN

One day I looked out through the wire fence that separated our front yard from the street and saw a boy approaching on a tricycle. He was dark-eyed and solemn, just starting to grow out of his baby fat. He was gnawing on a large apple.

"What's your name?" he asked, staring at me.

I stared back. "Weez," I replied. I continued to stare at him while climbing higher on the fence to get a better look. There was something odd about the way he spoke. And he was the first boy of about my age I had ever seen in Hilt. I was interested.

He climbed off the tricycle and approached the fence, making a jump for the top board. Effortlessly, he toed the wires and straddled the summit of the fence. He held out the apple to me.

"I'm Robbie," he said. "Want a bit of ahple'?" I nodded and took it, fascinated, and we spent a blissful half-hour sharing fruit and spit before he climbed down and pedaled away.

Robbie lived two doors south of us on Front Street, with his parents and his little brother Stevie, who was Elizabeth's age. He was four years old, and I was three.

His mother was a war bride from Manchester in England. Her boys said "*to-mah-to*" and "ah-ple." She had black hair and the kind of milk-and-roses complexion peculiar to damp countries in northern Europe.

Her husband, Sam, was a lean man who could fix bicycles.

The boys were wiry creatures with olive skin and warm brown eyes. Robbie was courteous, solemn, imaginative, and logical. His brother Stevie was Elizabeth's age, and in the years to come he would often flee screaming from our yard after a brawl with my sister, who gave as good as she got. When Elizabeth in her turn came howling from Stevie's yard, Grandmother dabbed alcohol on her scratches and told her that if they were just going to fight, she might as well stay home.

To the north, separated from our yard by a narrow alley, lived the only other girl child on our street, Caroline Ruger. Her parents had married relatively late in life, and she was their only child. Two years my senior, she towered over me—a pale, slender blonde who usually thought me too much of an infant to play with. Until I was five, our relationship consisted of insults shouted over the fence. She demanded to know where my father was, and when I could give no satisfactory answer, she jeered that I didn't have one. Our exchanges usually ended when I fled into the house in frustrated anger and flung myself at Grandmother's broad, aproned waist. "Don't talk to her, then," was Grandmother's unhelpful advice. "Go play in the backyard by yourself." Muttering, I wandered out the back door, inventing comebacks to Caroline's insults. But by the time I returned to the front yard with a cap pistol and a head full of repartee, she had grown bored with taunting me and simply declined to appear.

When Grandmother gave a party for Elizabeth's third birthday, she invited Robbie, Stevie, and Caroline. After that, Caroline sometimes unbent enough to invite me over to her yard to play, probably encouraged by her mother. She spent most of these occasions showing me her vast collection of dolls and comparing mine unfavorably with them. On one visit, I fell from the picnic table in the Rugers' grape arbor, striking the back of my skull on the stone-flagged floor, knocking myself briefly unconscious. Shortly after that, the Rugers moved to Ashland. For years afterward, I thought my accident had something to do with their departure.

The summer I turned six, the alley over which Caroline and I had shouted was plowed up, covered with a thin layer of topsoil, planted to grass, and a new fence built in the middle. Landscapes changed, I discovered. People left, and people arrived.

A family named Fremd moved into Rugers' house, bringing two pale, twiggy daughters who played with dolls and tea sets, wore dresses all of the time, and clung to each other and howled when I threw dirt clods over the fence at them. They ran inside, while their mother glared at me through lace curtains. Even Elizabeth could scare these kids. Learning this, we ignored them and went back to playing with Robbie and Stevie.

Robbie was my first date. I remember the spring morning he stood on our porch, dressed in corduroy pants, brown hair slicked down, plaid shirt buttoned up, red poplin jacket zipped against the cool sunshine. He carried a square lunch pail on which Roy Rogers bestrode a rearing Trigger. He must have looked like a bread advertisement, but to me, on that day, he was the last word in sophistication. He was in first grade, it was Visitor's Day at school, and I was his guest.

That morning Grandmother had stood me up in the bathroom, brushed my hair until it crackled, then parted it firmly in the middle and yanked it into four tight braids—two big ones at the sides, and two small ones that pulled back the hair above my ears. She pulled the smaller braids back and yoked them together with a rubber band, then incorporated their ends into the two larger braids. Every follicle on my head screamed in protest.

I was wild to get into that school. Hilt had no kindergarten, and although Grandmother had been teaching me the alphabet using an old primer, it wasn't the same. I had never seen the inside of the school and was so excited I could hardly breathe, and I certainly could not sit still. Grandmother handed me out the door and into Robbie's keeping, visibly relieved.

Robbie clutched my sweaty palm as we walked down the sidewalk to the store, turned left, and headed up the long row of houses that ended just below Watertank Hill on the north side of town. Here the sidewalks were narrow, the yards shadowed by trees, the air cool on my damp plaited hair. We cut down the side street that ran past Mrs. Sife's garden, crossed another street, and finally passed through a tall gate into the schoolyard.

The wood-frame school was painted a peculiar pinkish tan and stood on a knoll at the eastern edge of town, surrounded by box elder and juniper trees and a pale attempt at a lawn. Behind it, a sweep of rock-strewn rangeland climbed toward Skunk's Peak. Robbie and I walked into the south entrance of the building, and I stared down the long, dark, wainscoted hall filled with large, loud older kids. I smelled the heady combination of old wood, chalk dust, peanut butter, disinfectant, pencils, and dirty sneakers. I was taut with excitement.

Gripping my hand firmly, Robbie led me into a big room at the end of the hall. Light streamed in from a row of windows set high on the east wall. The west wall was a vast expanse of blackboard, smoothly shammied this Monday morning. Under it sat a large desk, and behind the desk sat an immense woman. She wore a flowered dress with a wide belt. She had soft brown hair and a mole the size of a grape on one pink cheek. Robbie led me up to this monster and introduced me. She smiled at me, and the mole moved over her glowing skin. Mrs. Davenport's voice was kind and soft, with a Southern lilt. She found me a small chair and set it beside Robbie's desk.

We sat on the first graders' side of the big classroom, beneath the stage, for the room had originally been intended as an auditorium. Across the room lurked the second graders. A few of the big girls pointed at me and giggled, but most ignored our side of the room entirely. While Mrs. Davenport kept me supplied with worksheets to color with Robbie's crayons, I kept one eye on the second graders, as Mrs. Davenport led them through demonstrations of unimaginable knowledge and skill at the blackboard. Why, they were learning to write in *cursive!* I could barely print the alphabet. I was, however, relieved to see that I could read what Mrs. Davenport wrote on the blackboard.

At lunchtime, Robbie walked me down the wide hall and out the school's big, double front doors. We sat on the steps below the wide porch and ate sandwiches and cookies and apples, sliced and peeled the way I liked them. Around us, big kids ran up and down the scuffed steps, laughing and eating. They frightened me, with their nearly adult size and aura of dangerous unpredictability. Even Robbie—brave, sensible Robbie—moved over to let them go by.

I recognized Larry Black, an eighth grader who lived across the alley from Robbie and who told us that the holes dug in the old playground behind that alley, beneath the ancient swings and teeter-totter, were devil holes. Step in one, he assured us, and Satan himself would grab our feet and pull us down to hell. We saw no reason to doubt him.

Robbie was courageous but quiet and polite. I could not have had a better first best friend. He possessed an innate streak of considerateness, foreign to every other child and to most of the adults I later met. He had the patience to explain things to me and to Stevie and Elizabeth.

I would do anything he told me to do, up to and including strapping on my toy six-guns and leaping off the woodshed roof to attack the Apaches, serene in the faith that anywhere Robbie went, I could follow. I sailed over the roof edge and landed on my stomach on top of one of the pistols. My caballero stayed just long enough to make sure I wasn't dead, then prudently ran for home before my howls could bring Grandmother running. He was gallant, but he was no fool.

Robbie was my introduction to male anatomy, as Stevie was Elizabeth's. I followed Robbie even unto his bathroom and waited and watched as urine thundered into the bowl. Peggy never tried to run me out, perhaps preferring that no fuss be made over girls being present while her boys spent a penny. More probably, she simply couldn't keep track of us all the time, especially considering the floor plan of her house. Its first inhabitant had been an amateur cabinetmaker who built a breakfast nook—one of only two in town that I ever saw—between the kitchen and the front room. And if you walked into the back of the closet in Robbie and Stevie's bedroom, you entered a passageway that emerged into the pantry behind the kitchen. A trip to Robbie's house meant a walk through the passage, one of many rituals in our lives, like never missing an opportunity to run through the dark and creepy passages behind the stage at the school.

In our games, Robbie was John Wayne, Errol Flynn, and Cecil B. DeMille rolled into one. His specialty was death scenes. "You go on, pal," he would choke out, grinding his sneakered heels into the dust, arching his back and clutching his flat stomach, gun dangling from nerveless fingers. "I'm done for."

The supporting cast—Elizabeth, Stevie, myself, plus Dwayne Dragoo and Edrith Cain, from the next street behind the alley—looked on, moved and awed by what we instinctively knew was a great performance. In our continuing dramas, Robbie was writer, director, leading man, and set decorator. A host of dramas played out in the yards and alleys and sidewalks. Robbie and Dwayne, as the oldest boys, played the leads and used all the good lines. The rest of us were shot and stabbed and ordered about. Robbie's superior years and command of logic settled most rebellions by the extras. I, for instance, always wanted to be a horse.

"Look, Louise," Robbie addressed me patiently, after I had insisted for the hundredth time on being Trigger or Silver or Fury. "If we had a horse, we'd use one, but we don't. Somebody's got to be an Indian. Now take the jump rope out of your teeth." I had to content myself with using Stevie's stick horse, or driving my sister down the sidewalk with a length of clothesline looped around her waist.

Mostly, we played Cowboys and Indians. It was a measure of our isolation that we had no idea that we lived in the actual West. We thought of the West and its nineteenth-century adventures as existing in some other place—not here, not on the California-Oregon border. In television Westerns, cowboys talked about Laramie and Dodge City and places called Dead Man's Gulch. They didn't talk about Yreka and Hilt. We certainly didn't think of ourselves as Westerners. We seldom saw a live horse, and we did not equate the re and white Herefords grazing through Hilt's streets in the springtime with the red-eyed longhorns of the great cattle drives. Yet we lived within a few miles of old men and women who had ridden in stagecoaches, seen bad men lynched, and remembered the days when Captain jack and his Modoc warriors fought the U.S. Army to a standstill.

When Robbie was seven, Peggy produced another child named Michael, or, as we always referred to him, The Baby. His arrival triggered in Robbie a new solemnity that increased during a summer in which he seemed to grow taller every day, making me seem to shrink in comparison. One summer morning I ran up onto his porch and knocked on the screen door, only to be met, after a long wait, by a subdued Robbie, who hissed at me, "You woke up The Baby! Mum's mad." I slunk down the steps and sat under the weeping willow tree on the front lawn. After a while I wandered home.

If The Baby engendered in Robbie a new sense of responsibility, Stevie spent more time than ever in our yard, fighting with Elizabeth. Robbie did not seem to miss him. On warm days, as Michael drooled in his baby buggy on the Flyte's front porch, Robbie seemed to actually *like* sitting beside him, reading a comic book and jiggling the buggy with one foot. I thought his attitude maddeningly superior and annoying. He didn't want to *do* anything except baby-sit and go to Cub Scout meetings with his new (male) friends from across town. When he came back from them, in his blue and gold uniform, he didn't want to play with us, and he wouldn't take me to meetings, because I was a girl, and too young besides. Somehow, he just wasn't much fun, anymore.

Two doors west of our house stood the largest dwelling in town, where the Company's resident manager lived. It alone of all the Company houses was painted. It had clapboard siding and a wide, roofed veranda running around three sides. Inside, a long entrance hall opened into a parlor with a fireplace and built-in bookcases. When I began reading books set in English country houses, I always imagined the characters taking tea in that room. Large windows looked out on Front Street through the branches of maple and horse chestnut trees. In summer, their branches threw the whole sidewalk into dark, delicious shade.

Old Mr. Baumann held court in that parlor on New Year's Day, backed by a blazing fireplace and a Christmas tree that touched the ceiling. He sat in a wing chair in his three-piece suit, his cane between his knees, a gold watch chain stretched across his ample belly. Mrs. Baumann served cookies and punch to the children, Tom and Jerrys to the adults. My sister and I sat on a real leather couch in our best party dresses, white ankle socks, and black patent leather shoes with straps across the insteps. We behaved ourselves on these occasions, to Mother's relief. At home, our table manners could be both colorful and inventive.

Our presence at these open houses was a reflection of our anomalous position in Hilt society. Divorce or no divorce, Mother belonged to the world of Front Street, to the society of the bosses. Her father was a foreman, which made him, as far as the Company was concerned, a gentleman. Grandmother and Aunt Jo and Mother were therefore ladies. Mrs. Baumann simply ignored The Divorce, although others did not.

The effect of The Divorce was greatest in the first few years after we came to Hilt. It set the entire family apart. Grandmother had never been especially fond of the bridge parties and teas organized by Hilt's clubwomen, but now she avoided them altogether. She visited with

Mrs. Sife and Mabel Hall because they could be counted on not to make curious inquiries about The Divorce. After Mrs. Ohlund publicly snubbed Mother, Grandmother never drank morning coffee with her again, pleading her grandchildren as an excuse. She still talked to her over the back fence, but she would never enter her house again, nor did Emma come to call anymore.

Another Front Street neighbor was important simply because of his effect on Hilt's public health. Louie Alphonse lived across the street from the Company office and was the plant's water, steam, and electrical specialist. He was also supposed to add chlorine to the town's water supply once a month, but sometimes he forgot. The next month, after everyone had enjoyed a bout of diarrhea, he put in twice as much, and the fumes boiled out of the taps and killed goldfish all over town.

At Christmas, Louie and his two unmarried daughters put up the town's best decorations, topped by a painted plywood sleigh in which a leering Santa careened across the roof, towed by a breed of reindeer never seen in Lapland. Lights ran all the way around the front porch eaves and the sides of the house. And on Halloween, the Alphonse sisters handed out the best candy bars.

Louie's daughters both worked in the Company office, and it was understood, somehow, that they always would. Elizabeth and I were somewhat in awe of them, for they had had considerable experience of death. It set them apart, as The Divorce did us. Louie was a widower, and his daughters kept house for him and presumably consoled him for the loss of his only son in World War II.

"Not much of a loss, if you ask me," Mother said one day when I pointed to the name "Alphonse" carved on the War Memorial in front of the Community Church. I stared at her in surprise, shocked at this affront to the holiness of the dead. Until then, it had never occurred to me that anything bad could be said about them.

CHAPTER 5

Father, Mother, and Me,
Sister and Auntie say
All the people like us are We,
And everyone else is They.
RUDYARD KIPLING

Mother was thirteen years older than her sister Josephine, a gulf of time that placed them on opposite ends of the Great Depression. They came, in effect, from different generations.

Mother was born in Susanville, California, in 1923, a year after Bill Roush married the Danish girl he had met at the armory dances in Iowa. In Susanville, Bill worked on the greenchain at Fruit Growers' mill, where he quickly grasped the basics of lumber grading. He was, for the time and place, an educated man. Martha had liked that about him. The farmers who slicked down their hair and came to those dances only conversed about crops; Billy liked to talk about the world and politics. He knew where Denmark was. So it bothered her, after they were married, to discover that he seemed to have little ambition. She had a great deal of it.

A handful of photos remain from Mother's life in Susanville: Barbie, my mother, sitting on the lawn of a frame house, her hair bobbed like Christopher Robin's; her mother Martha leaning on a porch railing above a yard piled with snow; Bill and Martha and Barbie, picnicking near Mount Lassen, leaning against a Model A Ford. Susanville was not a company town, although Fruit Growers Supply Company owned some of the houses. The county seat of Lassen County, it was big enough to have an existence apart from any one lumber mill; big enough to survive on ranching and tourists and county government, even as the open forests of ponderosa pine on the dry volcanic plateau were high-graded by successive waves of logging.

Fruit Growers actually began its Susanville operations years after it had stumbled into the lumber business at Hilt. Owning its own lumber facilities turned out to be so profitable that eventually Fruit Growers was to own three mills—at Hilt, Susanville, and nearby Westwood.

At Susanville the landscape spills out into a series of escarpments and valleys, onto the cold plateau between the Sierra Nevada and the southern beginnings of the Cascade Range, as if the land itself cannot decide how it wants to look, what it wants to be. Out on the prehistoric playa of Honey Lake, the April winds blowing off the Sierras at thirty miles an hour picked up the alkali dust and sent it boiling into town, where it mingled with the mill smoke and floated down onto cars and furniture and children. When Barbie was five years old, the spring dust storms coincided with an outbreak of meningitis. She came home from kindergarten with a feverish sickness that responded to neither of Martha's standard remedies: Capsoleum rubbed on the chest; and toddies of lemon juice, hot water, honey, and whiskey. Martha took Barbie to a doctor who listened to her chest and felt her throat glands and told Martha to take her home and put her to bed and wait. Barbie grew worse, running a high fever, and the doctor came to the house, and stayed for a while. Yes, he said. Meningitis.

There were no antibiotics, no serums. There was nursing and waiting, until one night Bill and Martha and the doctor all sat by Barbie's bed, waiting for the end.

Martha sat at the foot of the bed, believing that it was her fault that this child was going to die. She prayed, reflexively, for God to let Barbie live, but deep inside a small voice repeated that this death was God's way of telling her that she should never have left Iowa, should never have left her own mother, who needed her. This was about her, about her sins, her failings, and God was taking Barbie away to punish her for them.

The men looked at the child and, in the way of men, were angry and helpless and afraid of their failure. The doctor had no more ideas, no more tools. He picked up the twig-thin wrist one more time and counted the thready pulse, while the parents, their faith in medicine waning in the face of its uselessness, watched him, expecting . . . what? A miracle?

The pulse was weaker, now, and the doctor knew the end was not far off. He remembered a remedy that sometimes worked in moribund cases. He told Martha to go into the kitchen and make a pot of very strong coffee. Bill looked at the doctor, believing he was only giving her something to do, a reason to leave the room while her child died. But then the doctor rummaged in his case and pulled out an enema bag. When the coffee was done, he told Martha to pour it back and forth between two pans until it was blood-warm. Then he rolled the little body over on its side while Martha tucked piles of towels beneath her, and gave Barbie a series of coffee enemas, until the pot was gone. When he was finished, they laid her on her back, and tucked the blankets around her, and waited while the doctor, his bag packed up once again, stared at his watch.

When he took Barbie's pulse again, a flicker of surprise crossed his face. He pulled up her eyelids, watched her chest rise and fall, and finally pulled his stethoscope out and listened to her chest. He turned to Bill and Martha, his long teeth beginning to show beneath his old-fashioned gray mustache.

"By golly," he said softly. "I think she's going to make it."

In the spring of 1931 the Great Depression was so bad in northern California that Fruit Growers was laying off workers at the Susanville plant, Bill among them. He and Martha sold their furniture, loaded the rest of their possessions into the Model A, and drove north to Mount Shasta City. They moved into one wing of Grover and Katie's big two-story house on Chestnut Street, and Barbie went to school with her cousin Katherine. The mills around Mount Shasta were working only a few days a week, and none were hiring.

In July, all of Bill's brothers and sisters and their families met in Rainier for a family reunion with Mother Lena. Barbie turned eight years old there, and her grandmother gave her a birthday party. Then, from some hoard of her own, the widowed Lena excavated five hundred dollars and loaned it to Bill. Perhaps he and Martha could find work back in Iowa. At least, they could visit with another set of relatives for a while.

Niels and Anna Dittmar had lost everything but their house in the bank failures that swept the Midwest. They rented the house out to Martha's brother Andy and his family—a wife named Edith and four children. Niels and Anna moved in with their daughter Sophie and her husband, a taciturn man named Schroeder who had a farm. Schroeder shouted at Sophie, worked his draft horses into a quivering lather, and alternately ignored and bullied his small son. But on the farm, there was work and lodging and food, and so Bill and Martha joined her parents there, and in the fall, Barbie started going to the small country school a mile away.

There was no work off the farm for the men, but Martha found work in a cracker factory in Clinton. She boarded in town with Andy and Edith during the week and immediately remembered why she had never liked her sister-in-law. She was relieved when the cracker factory closed in November, although back at the farm she found a husband on the edge of revolt. He had taken about all he could from Schroeder, whose treatment of his unpaid help was only a shade better than his treatment of his horses, his son, or his wife.

Bill wrote a letter to his old manager in Susanville, asking about work. Mr. Birmingham wrote back that although there was still no work in Susanville, the mill up at Hilt might be hiring a few experienced men. He enclosed a letter of recommendation, and when Bill duly sent that letter on to Hilt, he received a job offer in December. He waltzed Martha around the room, singing in his bass voice.

Little remained of the loan from Bill's mother. On the journey back to California, Bill and Martha slept in the car, with Barbie bundled between them and all the coats and blankets piled on top. On a bitter clear evening they coasted down into Salt Lake City, Utah, where Barbie, pressing her nose against the freezing windows, watched the moon come up over the Wasatch Range, turning the world to silver. She asked her mother if Santa Claus would find them. Sixty years later, she chided herself. "Dumb kid," she laughed. "Always thinking of myself."

Christmas Day arrived in Elko, Nevada, in the middle of a blizzard. They found a store with a gas pump, where the storekeeper unearthed a sack of candy for Barbie, and let the travelers stand by his stove for two days, until the roads were open again. Santa left a doll for Barbie on the back seat of the Model A.

Bill started work at Hilt in January, working three days a week at the box factory. Martha and Barbie spent their days a few miles south of town in one of a row of auto cabins, for there were no houses available at Hilt. Barbie walked to nearby Sheldon Rock School and finished the second grade there. In the summer of 1932, in time for Barbie's ninth birthday, a house came vacant in Hilt. Bill and Martha gathered together a few pieces of used furniture and moved in.

If the Company was giving him a house, Bill figured, then the long-rumored closure of Hilt must be only a rumor and not yet a fact. Then the New Deal brought the Civilian Conservation Corps to a new camp ear Hilt, which in turn bought lumber from Fruit Growers. When the CCC boys began building a road west from Hilt into the Beaver Creek drainage, the Company began to get the picture. The new road meant that the Company could enter the brave new world of truck logging.

But even as new life was pumped into the town's industry, the next four years saw the Depression grow deeper and become a habit. Martha watched as her husband seemed content to spend his days as a box factory hand, and she did not like it. Her mother's letters from Clinton now told of better times back east, of bricks and mortar newly laid at WPA projects, where the Dittmar men found work.

The life of factory work and cheap rent that meant security to Bill was, to Martha, just a stopping place on the way to something better. Barbie would—must—have an education, must travel beyond this world of logs and smoke, must go to college. Perhaps, Martha thought, Bill was too much like his father, too beaten down by the fear of destitution, and she would have to push him, would have to make herself into another Lena. She was willing to do that, for Barbie.

Martha heard of a corner grocery store in Medford, owned by an elderly couple who wanted to sell out. She added up all their savings and figured that a down payment was just possible. By opening a bakery and doing the baking herself, selling coffee and pastries in the morning, they could draw extra trade. The store had survived the hard times well, and people always had to eat. But Bill remembered all too well, he said, how his parents had struggled with the candy store. Why should he give up a good steady job for the uncertainties of a business? It would mean borrowing money, too, and that terrified him.

"All right," Martha said, standing up over her ledgers and bank passbooks, staring at him. "You have to get a better job, then. I'm tired of living from paycheck to paycheck. If you're too timid to go into business for yourself, fine. But I want more for Barbie. She's going to graduate from high school and go to college. If you can't better yourself so she can do that-," she paused and drew a breath, "then I'll leave you, and take Barbie with me, and go back to Iowa. Mama and Papa are living in their own house again, and Papa and Andy have work, now."

Bill looked at the table, rose, and walked away without a word. For once, Martha made no attempt to go after him or placate him. He did not look at her or speak to her for several days. She cooked food and put it in front of him and refused to beg for forgiveness, as she usually did when he sulked. Then he began to bring hooks home—mathematics texts and manuals on lumber grading and measurement—and to immerse himself in them on nights and weekends. The box factory foreman was going to leave .in six months, and everybody in Hilt knew it. Bill knew that if he could demonstrate skill in lumber measurement and practical math, he had a good shot at the foreman's job.

By 1935, the job was his, and in triumph he moved Martha and Barbie into the house on Front Street that belonged, by custom that was by now as strong as a right, to the box factory foreman. And then in 1936 Barbie's little sister Jo was born.

At Josephine Edna's birth, most of the neighbors nodded sagely and said, "Change-of-life baby," but in fact the infant was planned. Barbie felt alone without sisters or brothers and envied the large families of her Italian classmates. True, the reproductive track record of Barbie's aunts and uncles would not have led anyone to predict that Martha would produce another child, thirteen years after Barbie's birth. Bill Roush and his six brothers and sisters would together produce a total of only five children.

In the twentieth century, Bill and Martha's generation abandoned the ways of their ancestors. They moved off the farms. Of Bill's siblings, Grover ran a saloon; Nick was a barber; Charlie drifted from one coal mine or logging camp to another; Fanny had married a lumber buyer; Mae and her husband ran a gas station; and no one seemed to know just what Frances's husband did for a living, except that it involved wearing a suit. Factory hands and barbers and saloon keepers couldn't afford large families. And their wives were not ashamed of their vacuum cleaners, washing machines, and Fords. They went to the movies and listened to the radio. They liked their tidy, well-vacuumed houses, far removed from dust and manure and all the inconveniences of the nineteenth century. In this new world, one or two children could have good clothes and go to high school and take piano lessons. Seven could not. And, for most of these women, one experience of the messy ordeal of parturition was quite enough. They thanked God and science for birth control, used it, and told horror stories about birthings gone wrong.

There was Fanny, for instance, who in an agony of modesty had hidden herself in the shrubbery when her water broke, lest the elderly doctor see her nakedness. By the time the doctor and a frantic Clare found her, the infant was dead. She never had another one.

Grover's wife Katie endured a long and painful delivery to produce Katherine, and when she saw the child's head, temporarily deformed by the birth, she shrieked and wept, and rather than go through *that* again, decided that one child was enough. Since her religion forbade birth control, Grover spent much of the next thirty years in his saloon and seldom came home for lunch.

Nick and his wife Hilma had stillborn twins and did not try again. Frances had no children at all, and Mae only one daughter, as did Charlie's wife Bertha.

But Martha had always wanted several children, and only lack of money had limited her family. As Bill's new job gradually increased their prosperity, Martha realized that it was feasible to raise a second child. If Barbie wanted a brother or sister, Martha would give her one while she still could. She tucked her diaphragm deep into her lingerie drawer and conceived again at the age of thirty-eight.

It was not an easy pregnancy. Her ankles swelled, she was constantly tired, and her teeth began to hurt. She drank quantities of goat's milk, which Dr. Langer told her was good for the baby's bones. She hated the taste, but religiously bought a gallon a week from Theo Avgeris, a

Greek immigrant who lived on the old Colestine Resort north of town. She choked it down, although the strong taste disgusted her, and the fat stuck to the roof of her mouth, cloying her taste buds for hours.

She went into labor one night in early September, feeling, oddly enough, better than she had in months, although she knew the infant was early. She woke Grandfather, dressed, and walked with him by the light of a flashlight out the back gate and down the alley, then east to the hospital. She sent Bill back home, telling him that it would be hours yet, and he might as well get some sleep. The nurse did not go next door and knock on Dr. Langer's door until after dawn. By nine o'clock, a tiny baby girl delivered with forceps was screaming in the next room. Martha was unconscious, her face swollen, and Dr. Langer fought to stop the bleeding as his patient's body was wracked with spasms. He had never lost a maternity case in this hospital, but he had never seen such a severe case of eclampsia, either.

High blood pressure, protein in the urine, edema: today, this fairly shouts a detaching placenta and a stressed fetus. But Doc Langer had never done a C-section in his life, and in any case the Hilt hospital didn't have the facilities for one. While Martha's kidneys dumped protein into her urine, while her red blood cells broke down and her blood refused to clot and her liver began to die, nothing was done. The baby, tiny and a few weeks premature, slowly starving of oxygen and nutrients, had to wait.

Barbie came home from the first day of school to find her father sitting at the dining-room table, his head in his hands. "Your mother's dying," he said bluntly, and she dropped her books on the couch and followed him out the back door, across the alley and the open space behind the garages, to the hospital. She looked down at Martha, a long lump of swollen flesh, like pale dough, in the metal-frame bed, deep in a sleep from which the doctor had been unable to rouse her. While Bill yelled at Dr. Langer, Barbie went into the nursery and stared at the red, misshapen head of her sister, wrapped in the blue receiving blanket Martha had brought to the hospital. The infant's eyes were screwed shut. Barbie had never imagined a human being could be so tiny, or so ugly. While she stared, her father's voice slowly receded down the hallway, following Dr. Langer, who refused to listen· to a shouting relative within earshot of the comatose patient. A nurse entered the room, thin and severe in starched whites, although Martha was the only patient in the hospital. She carried a baby bottle in one hand. "Here," she said, "let me show you how to hold your sister."

Martha remained in a coma for two days. When she awoke, Dr. Langer said she must have at least two months of complete rest. Martha thought it likely that she would have a great deal more than that, for when she came out of the coma, she was blind. She could see light and sometimes thought that she could make out shadows against the white walls, but for a week, as she lay in bed and groped her way to the bathroom and back, she knew despair.

Her sight gradually cleared during her three-week stay in the hospital. The Company hired another nurse to take care of the baby, after Bill demanded that Dr. Newton from Yreka come to Hilt to confirm Dr. Langer's diagnosis of eclampsia.

Two thousand miles away, Martha's parents read the telegrams from Bill and could not come. Sophie—tall, strong, cheerful Sophie—was dying of pneumonia. She had been sick with bronchitis for months, but hadn't gone to a doctor because Schroeder was always so angry about needless expenses. By the time Niels virtually kidnapped her from the farm and took her in to Clinton, she was delirious, and all Anna could do for her youngest daughter was to sit by her bedside and blame herself for not staying on the farm to help her.

No one told Martha when her sister died. No one wanted to upset a sick, blind woman. When she could see again, and Bill thought she was strong enough to bear it, he placed all the letters from Anna on her bed and went off to work. Martha read them, one by one. Emma Ohlund came in to check on her and found her crying silently, alone. "He killed her," she was sobbing. "Oh, he killed her, he killed her."

Martha's parents came out to California when the baby was three years old. They rode the train, two old people in comfortable, unfashionable clothes, seeing the West for the first time. In photos taken during that summer visit, Niels stands confidently in the back yard of the house on Front Street, legs braced, fists on hips, cap jaunty, anachronistic mustache gray but proud. Anna wears one of the long, loose dresses that old women still wore in the 1930s, women who had grown up wearing corsets and were now thankful to leave vanity behind. Her once-blond hair is almost white, and the most noticeable things about her are her delicate thinness and her light, hooded eyes. Beside her, Martha is taller and heavier, but mother and daughter both wear an expression compounded equally of sadness, resignation, and shy pride in the living children who stand beside them in the sunshine.

Martha's body crashed into menopause shortly after Jo's birth. Her liver walled off the damaged tissue and her blood pressure stabilized. Her body wanted to live, but her mind fell into a deep depression. Change of life, her doctor said. Just One of Those Things. No hormone pills, no Prozac, no therapy. Only letters to and from her mother, and one visit, one chance to ask Anna how it had been for her.

When Martha was strong enough to do housework again, her endurance was continually tested by the endless demands of two dependents: an infant and a husband who seemed to grow less mature as the years went by. Barbie, who had shouldered the housework while her mother was in the hospital, continued to tend the baby and run interference between her parents. When she entered high school, a year after Jo's birth, she sometimes came home to find her mother sitting in the twilight at the dining room table, an arrangement of playing cards laid out on the table before her.

Martha had learned how to tell fortunes from her mother. Although she never taught Barbie how to do it, believing it too connected to Old World superstitions, over the next several years Martha nevertheless read the cards more and more, when the unexplainable fears came, when she worried about her parents. Now in the near dark she shook her head at her daughter.

"It doesn't matter how many times I lay them out," she said sadly, "they always foretell change, and loss, and death. Something terrible is happening in the world."

"There's a war in Europe, Mama," Barbie would say, as she turned on the lights and the radio.

When Martha waved good-bye to Niels and Anna at the Hilt train depot in 1939, she knew she would never see them again. It made her feel dead inside—or something did, perhaps the same something that caused her to feel no desire for her husband anymore, that caused her to feel that she had forgotten why she had ever wanted him. "I wish," she told her daughter in the chill winter dusk, "that he'd just leave me alone," and sometimes Barbie was not sure whether Martha spoke to her or to a distant Anna.

What Barbie did know was that her mother's illness and depression had shaken Bill to the core. Having nearly lost his wife, he became more demanding of her, as though he blamed her for almost dying. His fear became possessiveness. Be there for me, always, his demands said. Never leave me again. Once, his wife had loved to dance and had enjoyed a few drinks at dinner parties. Now, she came home from the 1939 New Year's Eve dinner dance at Hilt's hotel only to stumble coming up the front steps and collapse on the living room floor. Barbie, waiting up for them, saw her father throw up his arms in hurt disgust.

"You're a drunk!" he yelled down at her. He swayed, whiskey on his words, as Barbie knelt beside her mother and told him firmly, "You go to bed. I'll take care of her."

Barbie got her mother up onto the couch, took off her shoes and stockings, and rolled down her girdle. She covered her with a blanket, found a pillow, and slid it under her head. She pulled out her hairpins. Then she sat down on the rocking chair and read, until she too fell asleep, her chemistry book fallen open in her lap.

CHAPTER 6

The lads that will die in their glory
and never be old.
A. E. HOUSMAN

Martha Roush may have been the only person in Hilt to whom the opening salvos of the Second World War came as a relief. Late in 1939 she began sending food packages to her cousins in Denmark. She joined the Red Cross and by 1940 was going to meetings sponsored by Civil Defense, which appointed a Civil Defense coordinator for every community. After Pearl Harbor, she joined the Aircraft Warning Service and took their course in aircraft identification. Less than a hundred miles from Hilt, a Japanese submarine had sunk an American merchant ship near Crescent City, and rumors of invasion shot like rockets up and down the coast. She ran outside at the sound of an airborne engine, binoculars around her neck, to search the sky. Since Hilt was on the main Pacific air routes, she got a lot more exercise and fresh air. She began to feel better.

She rummaged through the pantry and closets and pulled out battered aluminum pots and helped organize community aluminum drives and rubber drives and paper drives. She walked the neighborhoods on collection days, her leggy daughter Jo trotting behind her or helping to drag a child's wagon filled with scrap.

As loggers left the woods to join the services, the Company ran short of logs, even as demand increased for shipping boxes. By 1942, they had to cut back to running only one band saw in the mill, since there were not enough logs to run two saws. By 1943, they stopped trying to find contractors and hired a crew of their own loggers. They bought trucks and started hauling logs, as the independent truckers dropped out of the market, one by one. The Company bought more timberland on Indian Creek, west of Yreka. They built a logging camp and hauled logs into Yreka by truck, then loaded them onto rail cars and shipped them to Hilt.

During the war, Martha's parents died—her father of a sudden heart attack, her mother a year later of pneumonia. She could not go back to their funerals. At night she was haunted by their faces and by the photograph of their gravesite that her brother sent. She awoke after midnight and took her binoculars out onto the front porch and stood listening. Sometimes she heard a flight of bombers going over, flying south from Seattle to air bases in California. She looked across to the railroad depot where she had last seen her parents alive. She went back inside, unfolded a blanket and fell asleep on the couch, or stayed up to iron, one ear on the sky.

During the war, Bill spent more time than ever at the factory, trying to squeeze one more box frame out of a finite number of board feet, training new crews made up of women, who surprised him. They were good workers, quick to learn, their strength surprising. He didn't have to convince them he was the boss; they accepted it from the beginning.

Martha knew how easy it would be for her husband to have an affair with one or more of those women. She had heard the stories about how Things Happened. Rumors flew around town that at least one of the box factory women would Put Out for men on their coffee breaks, between the lumber piles. Art Hall told Mabel; Mabel told everyone. Bill was terrified that Martha would think he was one of those men. It never occurred to him that she might not care very much whether he was or not.

The war led to affairs and to unwise marriages and to odd attempts at creative free enterprise among people who had never displayed much of a bent for entrepreneurship before. A householder on Adobe Street, for example, started renting his two teenage daughters out to railroad workers on Saturday nights. The men would cut across the weedy fields, one by one, to the little house, after dark. When one of the girls came down with gonorrhea as a result of her contributions to raising morale on the home front, Dr. Langer told the Company, which shut the brothel down and told the father not to do that anymore. Perhaps they would have fired him, but there was a war on.

"Where in the world was the *mother?*" Martha demanded of the air.

"Well," answered Barbie, "everyone says she's not very smart." Running a whorehouse on the side may not have struck the Company as grounds for dismissal, but it took more seriously the complaints of his neighbors that he was a Nazi sympathizer. A native of Germany, he never made any secret of his admiration for Hitler, just as one of Hilt's oldest Italian residents infuriated his neighbors in Little Italy with his loud praise of Mussolini. But the sight of Mr. Goodpasture, the Nazi, raising and lowering the American flag every day in the center of town was evidently less palatable than the same chore performed by Goodpasture the Pimp. The Company appointed someone else to tend the flag.

Many of Hilt's young women began entering nursing schools. It was the romantic thing to do, if you couldn't be drafted. In 1942 Barbie entered the prenursing program at Southern Oregon State College in Ashland. Two years later she was studying nursing in a large teaching hospital in Portland. Nurses were snapped up by the armed forces as soon as they graduated. The hospital, having lost so many doctors and nurses to the war, responded by treating student nurses like slaves. Barbie walked an entire floor of patients during the night shift, then attended classes all day.

Late in 1944, those doctors who had so far managed to avoid the draft were surlier than ever with the students. One night, a resident who had just received his draft notice took his fear and anger out on the nearest target. He ordered Barbie to prepare the body of a dead patient for delivery to the morgue.

The corpse was that of a teenage boy, who had lingered in a coma for months following a headfirst dive into the shallow end of a swimming pool. Barbie, deathly tired, could have slipped away after the doctor left and found something else to do, leaving it for the day nurse. But she compressed her lips and began, only to discover that the body was already decomposing, the skin sloughing away at the touch of her gloved hand. When she attempted to clean out his mouth, his upper teeth dropped off into her hand. No one had suspected they were false.

For a couple of seconds, she thought his skull must have been falling apart. Shuddering, she finished the task, then ran to a sink to quietly retch up bile from her empty stomach. Back at the nurse's station, she wrote out a letter of resignation. By the end of the war, Barbie was a bank teller in Ashland, where the greatest hardships were a dearth of cocoa and silk stockings.

World War II was the defining, central myth of my early childhood. None of us remembered it, of course, but we lived with its results every day. On the front lawn of the little white community church, where we went to Sunday school and vacation Bible school, stood a granite memorial stone about five feet high. Its polished gray surface bore the engraved names of the men from Hilt who had died in the great conflict.

That list of names is remarkable for several reasons. Hilt only had about four hundred inhabitants. When a town that small gives ten young men to the body count of the twentieth century's bloodiest conflict, the result is that about ten percent of that town's draft-age male population has perished. That they died in such numbers is sheer bad luck; that they went in the first place was taken for granted. Siskiyou County boys enlisted by hundreds, by thousands. Some of them just didn't come home.

Three of Hilt's dead were from Italian families, some of whom had fled Italy after World War I to avoid conscription into Mussolini's new army. Their sons, however, went willingly into the American ranks, however unheroic or unmemorable their short civilian lives. Gordon Alphonse, for instance, the only son of his father, died piloting a fighter plane on a mission over Germany. To Barbie, he had been an arrogant kid who ordered his sisters about and changed his white shirts several times a day—shirts that his two sisters spent hours washing and ironing. But *they* were only girls, brought up to believe that brothers were more important.

Baumgartner died in Florida when his training parachute failed to open. Dunaway's P-38 went down over an island off the coast of New Guinea, as he tried to locate a Japanese submarine. Buster Nelson, a Hilt sailor aboard a nearby destroyer, saw him go in. Another Hilt boy, remembered as something of a clown in school, was kidding around with his squad on a quiet island in the South Pacific when he sat on a dud bomb, which turned out not to be. A letter assured his parents of a valiant death, but the unedited version of his fate made it back to Hilt through another Siskiyou County boy in his outfit.

Alphonse, Bernheisel, Clark. Dutro, Ladd, Baumgartner. Capello, Dunaway, Harris. Mottern. They died heroically and they died stupidly; they died in the Pacific and in Europe and in the States; they died one at a time, drop by drop, life by life. They were the brothers and sons who never came home. A coup le of them, perhaps, might have married the Alphonse sisters.

Or perhaps not. For most did come home. They married and sired the children who would play with Elizabeth and me in Hilt. Virtually all of our playmates' fathers had been in the war, which to us had ended about thirty minutes ago; the war was in the games we played and the slang we used, in our lungs as we pumped the swings higher and higher in the schoolyard, then leapt out at the high point of the arc, screaming, "Bombs away over Toke-ee-yooooh!"

When television came to our house, it brought not just *Howdy Doody* and *Dragnet* but dozens of war movies and documentaries, which reinforced the legends the fathers told. Sometimes, as we walked by the memorial stone, a kid would point to one of the names and say, with a mixture of sadness and pride, "That's my uncle." I wanted a dead uncle, too.

We had Grandfather, of course—on his dresser stood a tinted portrait of the Bill Roush that was, circa 1918. He wore a uniform the color of manure, the twin stripes of a corporal, and a magnificent head of thick brown hair. But he had never made it overseas—his greatest privation had been maneuvers in the southern California desert and getting separated from the trucks with the water but *not* from the trucks with the salt pork sandwiches. Besides, no Hilt kid bragged about a relative in World War I anymore.

Many years later, I learned that I had had a veteran uncle all the time—a live one. Uncle David had been a B-29 pilot, and really had let bombs fall away over Tokyo, on their way to incinerating Japanese children below. His plane was one of hundreds that roared over the battleship *Missouri* in September of 1945, while below them Japanese officials signed the surrender agreement and wished that they were dead. My father had been in the navy during the war, on an oiler in the Pacific, for a breathtakingly boring couple of years, during which the only excitement was round-the-clock kamikaze alerts during the Okinawa campaign. I knew that my father had been in the navy, for Mother had told me when I had asked. But she made it sound as though he spent most of his time in the brig, and I did not press her for details.

So I made use of what I had. My World War II heroes were the fathers of the kids down the street, across the alley, up the hill, and the unknown men on the granite slab.

On a large shale boulder perched on the thin soil of the big unkempt yard surrounding the old Club Hotel, generations of Hilt children had scraped their names with pocketknives. One day, crouched in the dirt behind it, shielding myself from a bitter, late winter wind, I noticed for the first time that one of those names matched a name on the churchyard monument. A real soldier had been here once. But he had not been a soldier then. He had been like me: he had played amid these rocks, heard the noon whistles blow, flown kites from Watertank Hill in the spring. He had caught frogs in the creeks and gathered pop bottles to trade in for the deposits. And he was utterly, irrevocably dead; not just fled into the incomprehensibility of adulthood, but just . . . gone.

We believed in ghosts, in those days. Suddenly, I didn't want to be on that hill anymore, under the flat gray sky that always came before a storm. I wanted to be home, drinking cocoa at Grandmother's house, in front of a warm TV.

The names on the monument were just names to me, the names of young, shouting, smoking men, unknowable. But the crude carving on the rock in the hotel yard was the name of a lost child, and that was different somehow. I did not want to think about him, and I put his name out of my mind. That night, it snowed, and in the morning whatever cold ghosts wandered that place were happy ones, riding fearless beside us as we careened our sleds down the steep little road beside the hotel.

CHAPTER 7

Oh may I squire you round the meads
  And pick you posies gay?
—'Twill do no harm to take my arm.
  'You may, young man, you may.'
A. E. HOUSMAN

In the summer of 1954, the Company carpenters built an addition onto the south side of our house. I remember Charlie Bloomingcamp and the other white-overalled Company carpenters placing the foundation blocks and later standing in the midst of the stringers, hammering. It seemed like such a small rectangle, until the flooring was laid and the framing began. Then it looked suddenly huge inside, with three large windows, the walls painted pink, a walk-in closet, its own bathroom, and the first shower stall I had ever seen.

Mother bought three single beds and three dressers. Aunt Jo moved her portable record player in and practiced dance routines on the wide polished wood floor. Elizabeth and I happily used our new beds as trampolines and hung our plaster ballerina plaques on our very own walls. And from the day we moved into our new room, my memories of helping Aunt Jo with the housework on summer mornings begin.

We made our beds and dusted furniture and shook the throw rugs out over the porch railings. We pushed the carpet sweeper over the patterned living room carpet. As we worked, Jo played records—Ravel's " Bolero" and Elvis Presley's latest. In the afternoons, to keep us quiet while Grandmother took her nap, she put our own records on for us—Humperdinck's "Hansel and Gretel," sung in English, which fascinated and terrified us with its images of parents who abandoned their children in witch-haunted woods. Later, there were sets of 45-RPM records narrating stories about Snow White and Cinderella and the Wizard of Oz. We listened to them over and over, until we could recite the dialogue and sing the songs from memory.

In the back bedroom next to the kitchen, the old upright piano, with its heavy carved legs and uncertain keys, still stood. Aunt Jo had learned to play on it, and Grandmother could pick out chords and tunes by ear. Aunt Jo had a gift for music, and she could read the mysterious circles and dots and lines of musical notation. She took private singing lessons, too. Once a week Grandfather drove her into Yreka to a shady house across from the city park, and we could hear her clear voice singing scales from across the street, ringing out from the open upstairs windows.

Jo had inherited her father's loud, penetrating Roush voice, but in Jo's soprano throat it took on tremendous range and power. It trilled, it vibrated, it carried. She bought the latest sheet music and taught us songs, so that Elizabeth and I could sing "Unchained Melody" and "Cherry Pink and Appleblossom White" and "Isle of Capri" and "Love is a Many-Splendored Thing" and "Heart and Soul" before we could read.

Jo's singing, dancing, and piano lessons had given her skill, but she also had the infectious enthusiasm of the born teacher and a knack for choreography. Although neither Elizabeth nor I displayed any talent for the piano, we could hit any note she played, had excellent memories for lyrics, and learned dance steps easily. When I turned six, we both started taking ballet lessons from Karen Adele, who came to Hilt several times a week to give lessons in tap, ballet, modern, and jazz dance to the children (and a few women) of Hilt. Jo perfected her modern dance steps with Karen and learned the nomenclature of ballet, adding it to the melange of foreign phrases with which she peppered her instructions to us. "First position, *then* third! *Capiche?* All right, get your *derrieres* in!"

I wonder now that between the ages of five and eight, I could step onto a stage before five hundred people and perform a complicated song and dance routine, with my sister or alone, and feel only confidence. I never had stage fright during Karen's big spring and full recitals, which were usually held in a large auditorium in Yreka or Weed. Grandmother sewed our costumes, Mother got us into them and made sure we had our props ready, and Aunt Jo touched a little pink lipstick to our mouths and instructed us to break a leg.

Aunt Jo was our big sister. She was young, only thirteen years my senior, and as gay as our mother was serious. At five feet two inches, she was the tallest woman in the house, with burnished, slightly auburn hair, big lively brown eyes, and a wide mouth, set off by a mole at one corner. She had the zaftig figure in style in the 1950s, when large bosoms and hips and narrow waists were the fashion. When she put on her black fishnet stockings and leotards and did a dance number with a top hat and cane she had picked up somewhere, her gorgeous legs looked a mile long, and she was suddenly the tallest woman in the world.

Alone of all the adults we knew, Jo seemed not to have forgotten what it was to be a child. She alone came out of the house after a snowstorm, wearing rubber boots and a scarf, to help us make snowmen and pull us around on our sleds. She taught us how to full backwards into the fresh snow and semaphore our arms and legs to make angels. She tromped out the ancient pattern for the game of Fox and Geese in the backyard and showed us how to play. On summer nights, she played Ollie Ollie Auction Free with us, throwing a ball back and forth over the roof of the little building behind Flyte's house, where the Boy Scouts met.

I remember sitting in the audience in the auditorium of Yreka High School, watching her graduate. She twinkled her fingers at us as she came back down the aisle in her cap and gown, clutching her diploma, before the endless, happy summer. But of course the summer did end. I remember how preternaturally quiet the house was that September, when Grandmother and Grandfather drove her down to San Jose and dropped her off at "approved student housing" near the state college. The house was so much bigger, so much emptier, after she left. We seemed to rattle around in it like peas.

I was bored in the long rainy afternoons that fall and winter and must have driven Grandmother crazy, for she walked up to the school and talked to Mrs. Davenport, who would have me in her first grade class next year.

"I think she's more than ready to learn to read, if she's reciting from memory like that," Mrs. Davenport said, as Grandmother described how I had several children's books memorized, knowing when to turn the pages, reciting them aloud to my sister. She gave Grandmother some old phonics books, and later Mother found some of her old school books in the closet, and they sat me down with lessons, five days a week. I didn't actually need much encouragement. Copying out the alphabet was like drawing; reciting the letters and their sounds was like learning a song. Grandmother and Mother had read to Elizabeth and me for as long as we could remember. Now, with Grandmother to coach me, breaking the alphabet's code came easily, and before long I could sit cross-legged on the floor in front of the glass-doored bookcase in the front room and pick out the titles on the spines: *Gone With the Wind,Anthony Adverse, Ways that are Dark, Green Laurels, The Light of Western Stars, The Outline of History.*

One evening a week, Mother took me to the branch library in the Warrens Building. I still remember the smell of that room, never quite duplicated: a combination of varnish and glue and print on paper, of floor wax and stamping ink. I remember how in winter the welcoming light shone out onto the red flagstones of the portico. I remember the solemn ceremony with which old Mrs. Mendes, the librarian, stamped the books and handed them to me with a smile and a whiff of talcum powder. In summer, we combined an afternoon walk to the library with a stop at the soda fountain just across the hall, but in winter, in the long mysterious evenings, we walked back across the railroad tracks in the dark, clutching our books, the three street lamps on Front Street looking very far away in the night, as our breath went up like smoke before us.

Jo went off to San Jose State College in the fall of 1954 because that was what Grandfather wanted for his youngest child. The only problem with his bright plans for Jo's future was just that: they were *his* plans, not Jo's. Jo didn't want to study, although no one could work harder than she to get a song or a dance just right. She wanted to *perform.*

"You can't live another person's life for them," Mother said to me years later. "When I went to nursing school, it was because my mother had wanted to be a nurse and never got the chance. I thought it was up to me to complete that dream for her. But I couldn't, because in the end I didn't really want it for myself. But Dad more or less forced Jo to go to college, because that was what *he* had wanted and couldn't have. He should have known it doesn't work like that."

Our aunt came home from college for the Christmas vacation bearing gifts for us and strange accoutrements; she lit sticks of incense and stuck them in a little iron burner. It smelled a lot better than Grandfather's cigarettes, but still he snorted and complained. "Damn chink stuff stinks," he said.

Jo bought a set of castanets and a pair of bongo drums, which we slapped while she performed flamenco-style dance steps, an artificial rose in her hair. She wore black leotards and wrap-around skirts with exotic patterns and sat on the floor in the full lotus position with her eyes closed, trails of incense snaking around her head, two small acolytes behind her. She began to talk about a classmate named Raoul, who was from Guam. His opinions on everything from Eisenhower to rock-and-roll became current around the house. "Sounds like a pinko to me," Grandfather snapped irritably behind his copy of *U.S. News and World Report.*

Nine months after matriculating at San Jose State, Jo brought home a report card so bad that it came with a diplomatic note from her advisor, suggesting that Jo might be happier taking a business course—somewhere else. She had done well in music classes, but a major in music meant two years of general education—science and math and history and literature, and in these Jo was interested not at all. Also, it became evident that she had not studied hard enough in high school to be prepared for them and didn't particularly care. At home for the summer, she threw herself into rehearsals for a community talent show and devised a complicated dance routine for a regional talent contest that would be broadcast live from Medford's only television station. And she kept up a voluminous correspondence with Raoul, not even bothering to hide his letters. This infuriated Grandfather, who blamed the boy for sabotaging Jo's—his—college career. He copied out the boy's address and wrote him an angry letter, threatening him with deportation unless he stopped writing to Jo.

Raoul, enjoying a summer on the mainland, wrote back to Jo. "Tell your old man," he said, "that he's an ignorant reactionary. Guam is an American Territory, and I'm a citizen. He can't have me deported." When Jo flippantly quoted this to Grandfather, he shouted and seethed and stalked out to mow the lawn, leaving Grandmother to drop tears onto the lunch dishes.

But uppity Guamanians would turn out to be the least of Grandfather's worries. In that long hot summer of 1955, Jo disappeared each weekend with a carload of her old high school classmates and their friends, on night-long trips that took them down to Dunsmuir near the southern county line, or "over the hill" to Ashland, racing other cars full of teenagers, trying to make the steep eighteen-mile trip in less than thirty minutes. She never told any of us where she went or what she did, but she gained a reputation for being "fast," and by August a box factory hand was bragging that Jo had put out for him in the back seat of his Chevy, a charge that Jo haughtily denied, but that Grandfather seemed to think eminently plausible. Over his dead body, he roared, would she go out again that Friday night.

"He seems to think I'm a whore," Jo said to Mother one morning, as they watched Grandfather's fedora disappear over the railroad tracks. "Well, all right. I'll be the best damn whore there is, then."

"You don't know what you're saying," Mother said, as she slipped into her office shoes. "Don't do anything foolish." And Jo smiled.

On summer evenings, our front yard was a gathering place for a familiar crowd of children and teenagers. The croquet course laid out on the front lawn was the nominal attraction for some, but there were frequent breaks while the players lounged over to the front fence and hollered at the cars that cruised slowly down the street and sometimes stopped. Sometimes someone brought a guitar and sat on the front porch steps, working through chords, or singing. Girls in pedal pushers and oversized men's shirts sat on the fence and swung their anklet-and-saddle-shoe-clad feet over Grandmother's tulips. Boys in tight white T-shirts with the sleeves rolled up over packs of cigarettes leaned on the croquet mallets and sometimes whipped a comb through the long sides of their hair, anointed with Wildroot Creme Oil.

As the dusk faded into dark, and the line of mountains on the horizon turned from blue to black with the first stars, the little groups on the lawn broke up. We gathered the balls and the mallets and put them back onto the croquet stand and carried it up onto the porch, away from the dew. One by one, the pony-tailed heads of the girls switched their way over to the cars lined up at the edge of the sidewalk and disappeared inside. Jo gave a hasty look over her shoulder and wiggled her fingers at Elizabeth and me, then slid onto a front seat as the car began to roll, soundlessly, two boys pushing it, pumping their jean-clad legs, until it began to roll downhill toward the firehouse. The engine turned over; the car turned the corner beyond Tallis's house with a crunch of gravel and was gone.

Inside, away from the mosquitoes, Grandmother gave Elizabeth and me two six-ounce bottles of Coca-Cola, and we hunkered down to watch the Friday night fights on television. Later, primed with sugar and caffeine, we stripped down to our underwear, put white socks on our fists, and danced about the living room in imitation of Sugar Ray Robinson or Rocky Marciano. Grandfather, deep inside the newspaper, paid no attention to us.

I remember that as the evening wore on, Grandmother looked worried. When Jo wasn't home by nine-thirty, we knew that she was in trouble. When she wasn't home by ten-thirty, we squirmed in a frenzy of anticipation. Somebody was Going To Get It, and it wasn't us. Mother tried to bundle us off to bed, even tucking us in and hearing our prayers before shutting the door to the front room, but we were having none of it. We slipped out of bed and silently opened the door a few inches, so we could sit on the floor and watch the beginnings of the late movie. No one could see us, and in time we swung the door open a little more. The grown-ups were awake, and all the lights were on.

We were drowsy and cold by the time the movie ended and Mother came to place us, unresisting, back in our beds. But when we heard a car pull up, we leaped for the open window and watched as Jo slipped out of a dark coupe and up the front walk. We crawled to the edge of the front room again and crouched down. I can see her still, as she came in through the big white door, her face flushed, her bright rayon scarf tied over her hair. She paused just inside the door and looked at Mother and Grandfather and Grandmother, lined up on the couch.

"What's everybody so quiet for?" she said, too quickly, too cheerfully. And the fireworks began.

We heard most of it, although Mother, out of patience, dragged us back to bed by the hand and shut the door so firmly that we knew better than to try to open it again. We couldn't have helped hearing, even had we been outside in the yard. Several times we heard Grandmother plead, "Billy, please—the neighbors!" and much later a muted rumble still issued from the back bedroom, where Grandfather lectured Jo for two hours, while she sat on her hands on the piano seat and, to keep from going crazy, pictured the sheet music stacked beneath her, inside the piano stool—pictured each song and played all the notes in her mind, over and over.

I don't know if Jo realized when Raoul stopped writing to her that Grandmother had anything to do with it. In Hilt, mail wasn't delivered. You walked down to the post office for it, and the one universally sacred custom in Hilt was that daily trip for the mail. At the post office, you touched base with the world and with all your neighbors.

Long ago, Hilt's post office had been a tiny place at the far end of the Company office and store building, with seventy-five lock boxes and a single window. In 1957, when the Company built a new office across the railroad tracks, the post office moved into the old Company office, where there was room for 150 new mailboxes with combination locks, two service windows, and the Company's gigantic old vault visible behind the postmistress' desk.

As the only local representative of the federal government, the postmistress was an Important Person. That Grandmother should take her courage in her hands, following that sleepless night, and beg Irene Tallis, the new postmistress, to put aside any letters from Raoul and give them into Grandmother's hands alone, was a measure of how badly frightened she was. Grandmother told Mother what she had done, and Mother shook her head in doubt. Irene's husband, Bill, was one of Mother's bosses—a sardonic man with a remarkable resemblance to Thomas Dewey, right down to the pencil mustache and the cigar clamped in place below it. He treated Mother with a mixture of condescension and polite, leering contempt that was hard for her to,hear. And Mother didn't trust Irene. What if she broke the confidentiality of her office and told her husband about those letters?

As it turned out, Grandmother might as well have saved herself the trouble, (or even as she tried to run interference between husband and daughter, *that* war was already being decided on a totally different front. From that night of shouting forward, Grandfather lost all influence over Jo's behavior, except the ability to make her do exactly the opposite of what he wanted. She ignored what came to be known as The Great Sermon as though it had never happened. Weekend after weekend, the young men in jeans, cuffs rolled up, lounged over our fence and roared away into the dusk, clouds of dust settling behind them in the red sunsets. Jo came home even later, once at four o'clock in the morning, only to find Grandfather still up, reading and waiting for her. "Are *you* drunk?" he demanded as she caught her toe on the doorsill. She tossed her chin up proudly. "Yes!" she retorted, and marched back to her room.

But with no money and no job, Jo was tied to her parents, no matter how much they might fight. Without resources or connections of any kind, her hopes for a career in show business were scarcely realistic. A winning performance in a talent show sponsored by Medford's only television station led to a solitary gig at the annual dinner of a Medford lodge, where she danced in her top hat and cane for a hundred men. The lodge never paid her the money they had promised her. And now her letters to Raoul, who had always understood her better than anyone, stopped coming, no matter how many she wrote to him.

Grandfather would not give her money to seek her fortune in Hollywood or at a drama school. It was business college in Medford, he said, or nothing. In the fall, she knew, the long evenings would end and she would go to Medford and learn shorthand and get a boring job somewhere, hut at least she and Grandfather would be shut of each other. Then in late September, even that option closed off for her.

Mother suspected before Grandmother did, when Jo stopped eating breakfast and nibbled on toast and grapefruit juice and slept until noon. But not until Jo came home late one night in silent hysterics, collapsing just inside the screen door, did she know for sure. Jo sat on the doorsill, her legs sprawled out over the threshold, the screen door bumping against her calves. Her makeup was streaked with tears. "Help me, Barbie," she said, "Please help me."

Mother knelt down and pulled her to her feet. "Come on inside," she whispered, "and I'll fix you a cup of tea," and suddenly Jo broke into peals of suppressed, sarcastic laughter, rising and falling, subsiding at last into helpless giggles. Mother almost literally dragged her into the kitchen, sat her down at the card table, and poured tea into her, while the story came out. They whispered, but the household was asleep, loud snores coming from behind the closed doors of the front bedroom.

No, he couldn't marry her. He was engaged to someone else, a Catholic girl in Dunsmuir. He was Catholic himself. His family would never forgive him. Can't you help me, Barbie? You're a nurse, aren't you? Why not?

Mother's horror at Jo's plea had nothing to do with revulsion against abortion. She would have done it in a minute if she had had the skill and the tools. But she knew what it entailed and knew that the Roushes probably didn't have either the money or the connections to obtain a safe one for Jo.

So she had to tell her sister that she couldn't help her, that she didn't know enough, and that it was dangerous when you didn't know enough. Watching Jo dissolve into despair at her feet, Mother could only hold her hand, send her off to bed, and promise to talk to Grandmother in the morning.

As the women stood around the stove the next day after Grandfather had gone to work and Mother had called in sick, the options looked no more promising. Jo *thought* she might be two months pregnant. "You know how irregular I am, Barbie, " she said. The two older women considered Jo's figure, her still-small waist.

"Are you *sure?*" Grandmother asked.

"Sort of," Jo shrugged. Sure enough to be afraid, sure enough to run through the list of her boyfriends in her head, none of them a likely candidate for a shotgun wedding. Sure enough to know that she didn't want to tell Grandfather.

In the end, the women decided to wait another few weeks and to pray for a miscarriage. By then, Jo's thickening body told the tale. Mother volunteered to bell the cat herself, while the other women were out of the house. But as she delivered the news and tensed herself for an explosion, Grandfather simply sat on the couch, his hands spread over the top of his bald head. He was silent for so long that Mother began to think her words hadn't registered and opened her mouth to repeat them.

"She's got bad blood," Grandfather announced to the floor, "I always knew it. It's the Scott blood, coming out in her. The Scotts were never any damn good." He stood up, not looking at Barbie, pulled his hat down off the hook by the front door and walked out, headed down to his office in the box factory, on a Sunday.

When he returned that evening, he said nothing to the women. It was Thursday before he spoke to any of them again. Then, while gobbling his lunch, he asked Grandmother to pass the canned salmon. Everyone breathed.

That weekend, Grandfather loaded Jo and Grandmother into the Oldsmobile and drove them down to a doctor in Chico. No one ever knew where he had gotten the doctor's name; perhaps Dr. Schlappi, the last Company doctor, now in private practice in Yreka, had given it to him. Of that long weekend without our grandparents, I remember bowls of popcorn and cups of cocoa in the cool October evenings and Mother singing all our favorite nursery rhymes to us in a vain effort to coax us into sleepiness. The travelers arrived home on Sunday night, and on Monday Mother called in sick again. Jo told her the story that morning, as Grandmother slept off the exhaustion and sorrow.

The doctor in Chico had examined her, but as soon as she took off her coat, he knew it was too late. He accepted referrals from trusted colleagues, but he was careful. He examined Jo and told her that he was sorry. "You're five months pregnant," he told her. "The baby's moving. I'd be taking too big a chance to help you have a miscarriage now."

He spoke to the three of them in his office, telling them all what he had told Jo. Grandfather stood up, his face red, his voice shaking. "You have to get rid of it!" he shouted. "I could lose my job!"

The doctor shook his head. "I can't do that, Mr. Roush," he said quietly. "Now we have to decide what's best for your daughter."

"To hell with that!" Grandfather screamed. "What about *me?*" In the end, the doctor ordered Grandfather out of the room. He spoke to Jo and Grandmother alone, asked them questions, and finally told them about a place called the Fairhaven Home for Unwed Mothers, in Sacramento. He had referred cases to them before, he said.

"So now I'm a 'case,' Barbie," Jo said at last. "But they can't take me until early December; there's a waiting list."

Mother looked at her sister's swelling breasts and belly. "We'll have to get you a new coat," she said. "A big one."

A couple of weeks later, as Jo walked into the Company store with Mother, feeling secure in the wide-cut wool garment, they met Bill Tallis coming out. Afterwards Mother swore that he had looked Jo up and down, his eyes lingering just a bit too long at her midsection.

"He knows," she said to Jo.

"How would he know?" Jo answered. "Anyhow, he always looks at me like that."

"He knows," Mother said bitterly. And if *he* knew, then who else knew, or guessed? Would the Company really fire Grandfather if that knowledge went beyond Bill Tallis? She did not know. What she did know was that something had changed in her sister during that weekend in Chico.

"You know what really hurts, Barbie?" Jo said to her one evening, as they walked the dark sidewalks of Hilt for the exercise the doctor had ordered. "It was the way Dad acted down at Chico. He didn't care at all about me, about my health. When the doctor told him I was too far along for it to be safe—it was like I wasn't even there, like I didn't even exist, like he didn't care whether I lived or died. All he was worried about was himself." It was a hard thing to wake up to, that knowledge.

Jo wouldn't have been the first Hilt girl to have a baby without a husband. Up on Adobe Street, Hilt's only known unwed mother lived in a silent purdah with her child and her Portuguese parents. Her treatment at the hands of Hilt's matrons made Mother's treatment as a divorcee look positively welcoming. But Marjorie's father was not a boss; he was a poor immigrant with no status to uphold. Jo's case was different. As far as Grandfather was concerned, she had nearly destroyed him and might yet. For how could the Company respect a boss whose underlings bragged about screwing his daughter, especially when the stupid girl got herself pregnant?

CHAPTER 8

> Mountains will crash down,
> Troll-women stumble,
> Men tread the road to Hel,
> Heaven's rent asunder.
> THE ELDER EDDA

When Grandfather walked into the box factory every morning, it was to face a competition with every other department of the plant. Each branch of the Company's Hilt operation was expected to make a profit, on its own. Frequently Grandfather walked back to his office after supper, or on a Sunday, to work through columns of figures with adding machine and pencil, figuring the number of slats and uprights needed for so many thousand boxes, trying to pare inputs and increase outputs. By the fall of 1955, he knew that his job hung by a slender thread, so slender that he feared that a scandal involving his younger daughter might snip that thread and send him falling into a destitute old age. For the Company no longer needed a box factory and could have done quite nicely without Bill Roush, thank you.

Although his skill at driving a bargain and stretching less into more had saved the Company a great deal of money over the years, it would have been hard to find a worker or a colleague who really liked Grandfather. "A hard man to work for," was the most charitable estimate. Bill could live with that—even consider it a compliment. He could live with the dislike of his employees, who, after all, had to obey him or quit. He could live with the dislike of fellow foremen like

Frank Graves. But he had managed to annoy some of Fruit Growers' executives, too, and that was more dangerous.

Mr. Utke and Mr. Thomas of the home office near Los Angeles both thought the box factory superfluous and its foreman too expensive in these days of good, sturdy, cardboard packing boxes, and Mr. Thomas was the resident manager's boss. Although Mr. Thomas had been kind to Barbie, giving her an easy shorthand test when she first started to work, kindness went only so far in business.

Grandfather's knowledge of the Hilt operation made his fears even more realistic. When Fruit Growers acquired the Hilt operation in 1910, it gave little thought to the future of the lumber business as a whole: it simply wanted to escape from the monopoly wielded by the Pine Box Association, which had cornered most of the West Coast's supply of box shook. When the Hilt mill rescued Fruit Growers from this dependency, they went a step further and acquired the mill at Susanville and finally, in 1944, bought out the Red River Company's interest in a lumber mill in Westwood. But Grandfather knew that the Company had plans to close the Westwood mill in 1956 and to move some of the Westwood foremen to Hilt: an engineer, an electrician, a machine shop foreman, and a construction and repair foreman.

The old problems of supply and demand had changed. When Grandfather first came to Hilt, the plant's output had been regulated by day-to-day demand, of which there was little enough in the depths of the Depression. Now, it was supply that drove the cold equations in a boom market. But supply was ultimately dependent upon the peculiarities of western land disposal back in 1868, when the federal government had granted the Southern Pacific Railroad every other section of land, in a checkerboard pattern, for over a score of miles on either side of the railroad tracks that now ran through Hilt.

Over the years, Southern Pacific had sold or traded many of those sections, and lumber companies like Fruit Growers gained legal title to lands that gold rush entrepreneurs like John Hilt simply used. In the northern half of the Beaver Creek watershed, west of Cottonwood Creek, Southern Pacific eventually exchanged about fourteen percent of its lands with the U.S. Forest Service. About 36 percent was purchased by Fruit Growers.

The Forest Service's presence in Siskiyou County began with the establishment of the Klamath National Forest in 1905. In 1912 the agency began a timber reconnaissance in the Beaver Creek drainage. Of its forty-six thousand acres, twenty-six thousand were in private hands, twenty thousand on the National Forest. The timber totaled an astounding 413,381,000 board feet—245,000,000 on the privately held lands and 167,465,000 on the public.

The Forest Service began selling timber to Fruit Growers. In 1918 the Company negotiated a twenty-two-million-board-foot timber sale from the Forest Service and took over three years to log it. They were in no hurry: the Forest Service had no other customers in the area for large sales. The Company had plenty of time to extend its logging railroads. In 1922, they undertook another large sale and were still logging it two years later. Usually, Fruit Growers simply waited until its logging crews approached the boundary of a National Forest section, then applied to the Forest Service to purchase a sale there. When lumber prices rose, all too briefly, in the 1920s, the Company bought millions of board feet of milled lumber from Long Bell Lumber in Weed, already cut into boards one-and-a-half-feet wide and ten or twelve feet long. Fruit Growers used this low-grade lumber as box shook and sold the more valuable clear lumber milled at Hilt on the open market.

In the 1930s, the Forest Service again surveyed the checkerboarded sections in the Beaver Creek drainage, this time with an eye to trading some of them. During the Depression, and for many years afterward, the Forest Service and Fruit Growers traded lands in this area on a three-for-one basis: for every three sections of logged-over Fruit Growers land, the Forest Service gave the Company one section of virgin timber. The land exchanges helped Fruit Growers survive the Depression. So did another government program, the Civilian Conservation Corps.

By 1932, the Company knew that the timber within reach of its logging railroads would run out within two years. Lumber prices were still falling, and the Depression showed no signs of ending. The Company couldn't afford to build new railroad lines, or even maintain the ones they had. Trying to salvage what it could from economic disaster, the Company spent the summers of 1932 and 1933 logging every stick of timber it could reach with the steam donkeys. Two six-hour shifts at the sawmill kept up with the extra logs. The Company hired more graders, worked a night shift at the planing mill, and put the box factory on overtime, hiring Grandfather in the process.

But luck was with them, after all. In 1933, the Civilian Conservation Corps established a camp near Hilt and began building a road from Cottonwood Creek up over Four Corners and down into Beaver Creek. Eventually, that road would link Hilt with the Klamath River Highway that crossed the mouth of Beaver Creek. The CCC also built the Siskiyou Crest Road and other spur roads connecting to the old logging railroad grades. Often, they were constructed right on top of them. Forty miles of railroad grade were eventually converted into thirty-six miles of logging roads. The Company began hiring logging trucks to haul logs out of the woods in 1934. Building spur roads off of the CCC roads was an expense, hut by contracting out the log hauling, Fruit Growers saved money it would otherwise have spent on maintaining a railroad crew. As the new roads allowed the Company to access previously unreachable stands of timber, Hilt was reprieved.

World War II increased the demand for wooden shipping boxes, even as the Company struggled with fuel and labor shortages. During those years, Fruit Growers clambered to the upper ranks of California lumber producers, a rank it improved on in the 1950s, when it became the state's third largest lumber producer—an impressive rank, when one remembers that in those days Pacific Lumber and the other great redwood manufacturers were slicing their way through the redwood forests of the coast, where a single giant tree could he turned into dozens of houses.

In the 1930s, the old-growth sugar and ponderosa pines were still the Company's preferred quarry. The sawmill was designed to accommodate logs more than 20 inches in diameter. In the 1950s, it was refurbished to cut dimensional lumber, now that the previously scorned Douglas-fir, white fir, red fir, and incense cedar were becoming valuable. The Company returned to areas it had highgraded in the 1920s and even earlier, logging them again. Even with the old-growth pine forests around Susanville and Westwood gone, the Company could afford to be optimistic. The lumber market seemed to have entered a permanent high.

In 1946, the Company realigned the lumberyards, making room for a new planing mill and a five-hundred-foot-long crane shed, completed in 1954. It was the largest structure ever built in Hilt. A big logging-truck repair shop and three more dry kilns were built. In 1950 a slicer plant to make box tops was set up. This worked poorly and closed after only a year, and its failure may have been laid at Grandfather's door. In any case, it 'Vas the last change the Company attempted in the box factory operations. All the other projects, and the transfer of employees from Westwood, focused on the production of lumber for the wide-open construction industry. Grandfather knew all this and knew, none better, that every moment he now worked was a gift, a gift endangered, as he saw it, by his own daughter.

So he wallowed in his own hurt and fear and sense of betrayal and never saw them coming, the two great distractions that would make Grandfather's job the very last thing on the Company's mind for the next couple of years.

CHAPTER 9

As in dark forests, measureless along
The crests of hills, a conflagration soars,
And the bright bed of fire glows for miles.
THE ILIAD

David Powers was the most phlegmatic child I ever met. He seemed to have been born with the soul of a forty-year-old accountant. I never saw him hurry, never saw him lose his temper. He accomplished everything easily, in a sort of trance, borne along on a slow, fluid economy of motion, like spilled syrup. He made good grades effortlessly and seemed perpetually and wryly amused about something too cool to share with the rest of us. He was a year ahead of me in school, so every second year we shared a classroom. His father was a Company electrician, his only sibling a brother who lounged through life even more slowly than David. The Powers family played pinochle together in the evenings. We knew this because David's revelations during show-and-tell always included a sidebar about how many hands of cards the family had played the night before.

David gained a lasting reputation for sangfroid when a group of younger children, passing by his house, saw a small fire blazing on the roof. "David! Your roof's on fire!" we screamed toward the top of his head one day, as he sat on an old couch on his front porch, behind the shading hop vines. David rose, marked the place in his book with one finger, strolled to the screen door, and announced in a laconic monotone, "Mom, Dad, the house is on fire."

While he strolled back, sat down, took his finger out of the book, and began to read again, we sat down on the sidewalk across the street, in front of Johnny Marin's house, and watched. As Mr. Powers came running out to turn on the garden hose, we heard Mrs. Powers shouting into the telephone for the volunteer fire department. Another day, another roof fire.

Despite many house fires over the years, Hilt didn't have an official volunteer fire department until after 1950, when Fruit Growers began to subsidize one. Its response time was phenomenal, considering that house fires in town required the firemen to race from the plant to the small firehouse at the end of Front Street, where the pumper truck was parked, then dash to the scene of the fire. It would have been more efficient to train neighboring housewives in the use of the truck, but no one thought of that. Still, because they were home during the day and most of the men weren't, women discovered most house fires and often put them out.

"Martha, your shingles are on fire," said Emma Ohlund one day to Grandmother, as they stood chatting over the backyard fence. And Grandmother turned on the hose and aimed the spray at the blaze until it hissed and died.

We lived in a world of wood, and we expected fire. Everyone in Hilt burned wood for heat. Everything in town was made of wood—houses, sidewalks, garages. Although most people in Hilt had electric cooking ranges by 1950, almost every kitchen also held the small stove known as a trash burner, with two round lids cut out of its cast-iron top, which could be lifted by a small, detach able metal handle. The stove's body was cast-iron, overlaid with thin steel coated in white enamel. A door on the front gave access to the firebox and the ash hopper beneath.

In our front room was a larger stove, also with a cast-iron body, but with a separate jacket of brown enameled steel and a large door on one end, into which big chunks of wood and bark were thrown. A damper on the stovepipe theoretically controlled the fire's intensity, but I remember winter mornings when the stovepipe glowed red and heated walls to the danger point. Wood stoves, and their unpredictable flues, were the most common cause of house fires in Hilt. The stovepipes had a cap to keep out rain and snow, but none were screened. Many roofs were peppered with patches of new yellow-pine shingles, covering the places where sparks had landed and burned.

By some miracle, no one ever died in a house fire in Hilt, but Grandmother had a horror of fire, and both she and Mother awoke at odd hours to sniff the air and walk through the house and peer at stovepipes and place their palms on the wainscoting. Frequently creosote, building up inside stovepipes, roared to life, sending gusts of flame and black smoke belching into the morning breeze. No one ever cleaned or brushed out stovepipes, although it was true that in those days before airtight, fuel-conserving stoves, less creosote accumulated. Flue fires were simply regarded as acts of God. Out in the woodshed attached to our house, a ladder led up to the attic space. From its top, you could look down the length of the attic and see the ascending stovepipes, innocent of double walls or insulation, rising through the dry forty-year-old boards of the house frame.

The Company delivered firewood to any house in town, as much as you wanted, anytime you wanted it. After the big planing mill was built, you could request either rough mill slabs cut from the outer portions of logs, or mill ends, trimmed from the dry planed lumber. The planed pieces were great for starting fires, and the mill ends were thick with hark that smoldered all night. We loved the mill ends because they made good building blocks.

The wood was delivered in dump truck loads directly behind the woodshed, and the wood tossed into the shed through a window-sized opening. We would clamber to the top of a newly delivered load and toss slabs aside until we had a cavity that extended all the way to the ground. These foxholes made good summer forts, for the piles were always left outside to dry before being thrown into the shed for the winter.

The really large town fires—the ones that took out entire rows of houses—lay in our unremembered past. After the great Sunkist Hotel fire, the Company built the new Club Hotel on a hill behind the Company store on the east side of town, safely removed from rows of houses. When a cookhouse fire torched a second hotel and some dormitories in 1925, the Company rebuilt the structures and placed the new cookhouse off by itself. They laid down a water line from the Company store to the north end of the bunkhouse, several hundred yards away, running under the alley behind Front Street. Midway down the alley, a small red pumphouse stored a long canvas hose that could be dragged out to shoot water on a house fire.

The cookhouse eventually became the Company carpentry shop, where men in white overalls hammered culverts together from the fragrant wood of incense cedar, ran planes over white pine drainboards, and repaired window frames. We rode our bicycles near the doorway just to smell the cedar sawdust.

That wood burns, and that it should burn, we never doubted. The tepee burner beside the sawmill poured smoke and cinders into the air day and night; the valley's wind patterns carried most of it away, but still we were surrounded by smoke, steeped in smoke, recognizing its kinds and gradations—the sharp sulfur tang of a lit match, the whiff of kindling beginning to snap, the late-night sweetness of a stoveful of bark on the coldest winter nights. We knew the dark pitchy scent of pine burning and the unforgettably joyful smell of smoke from wet Douglas fir sawdust, pouring white with moisture out of the tepee burner and curling down toward the ground, just before a big snowstorm. We knew the smell of weeds burning in the roadside ditches in spring, when the fire department set them afire, and the occasional hint, on a south wind, of burning tires and garbage from the town dump.

Outside of town, in the woods, we knew that fires also burned. North of Hilt, on a ridge above the old stage station, a forest fire had burned in 1895. A few snags from that fire still stood on the ridgeline, in the midst of a vast field of brush, all created when a donkey engine without a spark arrester set some logging slash afire near a railroad logging camp.

In the days of railroad logging, slash fires sometimes grew so large that the sawmill was shut down and the millworkers sent out to fight the fire. In 1920, a fire in Ditch Creek west of town almost consumed the box factory, before the ever-dependable three o'clock wind from Bear Canyon blew the fire back upon itself.

In the early 1920s, a fire started in Bushy Gulch northwest of Hornbrook and burned to the top of Cottonwood Mountain south of Hilt and over into Dutch Gulch. A fire on Soda and Smoky Creeks burned Bridge No. 5 in the woods west of Hilt, leaving logging engines stranded on both sides of the trestle.

In 1924 a fire on the west side of Cottonwood Creek burned Company timber, and the plant and woods operations shut down while workers fought the blaze.

In 1926 a fire started on the east side of Cottonwood Creek one Sunday afternoon and nearly burned the houses on Adobe Street. Mrs. Stonehouse, the resident manager's wife, formed a brigade of neighborhood children to slap out spot fires with wet sacks.

Before the white men came, fires burned in the forest every few years, cleaning the ground of dead limbs and duff. The Shasta people set fires to encourage the growth of basketry materials like bear grass and hazel, to keep the trails open, and to keep the poison oak at bay. When the yearly burnings ended, and the logging slash piled up, the formerly gentle fires turned large and fierce, so that first the loggers and then the Forest Service came to see fire as the enemy and put out every one they could. By the early 1950s, ground fires burned fewer acres, as lightning fires were swiftly attacked and suppressed with the newest tools of technology—smokejumpers, retardant bombers, fire engines, and even a few helicopters.

The Siskiyou Mountains share the Mediterranean climate of California and southern Oregon. Rains begin in late autumn and continue, with intervals of snow, until May. Almost no rain falls in the summer, and the Hilt area gets only about twenty inches of rain a year. On droughty south-facing slopes, trees in the old-growth forests were widely spaced because there simply was not enough soil moisture to sustain dense stands over a long period of time. Ground fires killed excess seedlings, so that during drought years the reduced numbers could survive. But logging removed the biggest, most fire-resistant trees, and fire suppression allowed species like white fir to gain a foothold on the logged slopes. By the 1950s, where once there had been eight or ten huge pines per acre, now fifty or a hundred young firs grew, changing a pattern of life and death hundreds of thousands of years old.

The Company and the Forest Service looked on the newer, thicker stands and saw a future timber resource, not a fire hazard. They were confident that a combination of vigilance and postwar technology would protect the forests from fire. On peaks all over the forest, men and women summered in lookout towers and scanned the ridges and canyons for smoke. Buckhorn, Deadwood, Hungry Creek, Dry Lake, Collins Creek Baldy—from these heights lookouts picked up lightning strikes and telephoned the Forest Service dispatchers and each other as they peered down into the long canyons of the upper Klamath River and its tributaries: into Beaver Creek, Horse Creek, Scott River, Humbug and Little Humbug Creeks, Lumgrey Creek, Buckhorn Creek, Cottonwood Creek, Barkhouse Creek, McKinney Creek, Dutch Creek.

After the war, with gasoline and tires once again available, Grandmother and Grandfather resumed fishing, and Grandmother learned to tie flies. I remember her as she fished, bringing in trout after trout, which wiggled frantically as she gently removed the barbed hook and measured the fish against the flattened palm of her hand, which was the exact legal size, six inches. She inspected their flanks for the ugly round wounds left on so many fish by the sharp mouths of lampreys. She slid the keepers into her wicker creel, seized her walking stick, and waded further into the river, the line snaking out into the riffles, flinging shards of sparkling water into the sun.

Sometimes Grandmother put her fly rod aside and brought out her newfangled spinning outfit, with its clacking reel and monofilament line, which Grandfather made fun of, until she began to catch more fish than he did. Then he got a spinning outfit of his own, a bigger one. In winter, Grandmother's big wet flies, wrapped in orange thread and trailing a hackle of polar bear hair, were deadly when attached to a flashing silver spoon. In summer, to her disgust, the lures sometimes attracted suckers, which she tossed back, disappointed, for they often weighed several pounds. We watched them drift downstream, their strange downcast mouths working, their beady, worried eyes staring back at us as they sank slowly away.

On Labor Day of 1955, we went fishing along the banks of the Klamath River, ten miles south of Hilt. Grandfather, his old sweat-stained fishing fedora on his head, quickly disappeared upstream. Grandmother fished closer to the car, while Elizabeth and I followed her along the bank. We were not fishing that day, since we inevitably jammed the tip of any fishing rod into the rocks, or ensnarled the line into huge balls. Aunt Jo, taking a break from painting her nails and listening to the car radio, patiently unwound them for us, exclaiming, "How the *hell* did you do that?" while we giggled, because usually nobody except Grandfather said "hell." Mother, luxuriating in the comfort of slacks and sneakers, sat on a fallen log under the shade of the cottonwoods and sketched.

Our river outings followed a pattern. Elizabeth and I trailed after Grandmother along the riverbank, searching for agates or the opalescent shells of freshwater mussels, or trying to catch the trout fry trapped in drying shoreline pools along with doomed tadpoles. We poked at decaying fish carcasses, followed the hops of mud-colored yellow-legged frogs, and returned at last to the car, hair sticky and shoes full of sand, which we invariably dumped out on the carpet.

On this day, the air was hot and dry and windy. Stiff breezes ruffled the river pools, blowing the fishing line back so that one could scarcely cast. Grandmother gave it up in midafternoon and gutted her catch on the shore with her little knife, as I watched. Then she sat with us under the cottonwoods, pulling off her waders and sitting barefooted, sipping a can of beer. We watched the thunderheads building up over the north and east. The air was cracklingly dry, the clouds an unhealthy purple beneath, and a taut, expectant feeling hung over the little grove. Grandmother began sucking her gold bridgework, and Grandfather came walking briskly back down the bar as the wind thrashed the trees and the first rumbles of thunder rolled down the brown hills.

We drove back to Hilt and unloaded the car while Grandmother made ham sandwiches, which by some sort of mutual consent we consumed on the front porch, facing southwest toward Cottonwood Peak and Sterling Mountain. Behind them, in the vast Beaver Creek drainage, lightning popped on the ridges, and not a drop of rain fell.

On Deadwood Lookout, south of the Klamath River, Orva Gosney saw it begin.

> The lightning started about 4:00 p.m. and it seemed to rain fire instead of water. Every strike started a fire. Two strikes came close to the tower, then about five strikes hit the bottom of Barkhouse Creek below us. I took a quick reading on them but before I could call the dispatcher the whole place exploded. A wall of fire roared

> up Haystack Butte. By 6:00 p.m. Barkhouse had a good start toward an inferno and the families on the creek were told to be ready to evacuate. The wind started to blow and the draft created by the fire, aided by wind, took sparks across the canyon to a brush field on Little Humbug Creek. Spot fires sprang up. Waves of smoke and fire rolled across the brush flat and each wave started fresh fires about two hundred feet or more ahead of the main fire. Wave after wave rolled toward the ridge and when they reached the top, they went over like a waterfall . . . it had jumped the river and was heading up Beaver Creek. The fire continued up Grouse Creek to a box canyon, lay dormant for a few hours, then with a terrific explosion that rocked my lookout and rattled the windows, accumulated gas ignited. Flaming debris shot hundreds of feet in the air and a crown fire was started in front of the lookout. By now, we were surrounded on three sides with fire and completely smoked in.

Hank Mostovoy, a Forest Service employee, saw the blowup on Barkhouse Creek, too. He saw the convection column, seventy thousand feet tall, that threw balls of flaming gas out to the side, which "ignited with a noise like dynamite exploding. One fiery ball jumped the Klamath River at Barkhouse Creek and detonated a mile away on the north side of the river. Beneath the smoke column was a tornado of fire, half a mile across, that pulled ponderosa pine trees a yard in diameter clean out of the ground, twisted them in half, and flung the pieces and the root wad fifteen feet away from where they had grown. The soil was melted and fused five inches deep. The firestorm burned nine thousand acres in an hour."

Called down from the lookout, Orva came back two days later, driving through walls of flaming snags. "The roar never let up. We were never sure which way it was going and expected to see it top the lookout ridge at any time." In October, rain slowed the fire down, "but until snowfall," she remembered, "snags and smoldering stumps in the canyons and ridges that had once been beautiful forests glowed below my lookout like tiny villages at night."

On that first evening, Clara Williams, the Forest Service fireguard on Hungry Creek Lookout, had just reported a smoke over the radio, when Fruit Growers' fuel dump on Sterling Mountain was struck by lightning. "A big ball of fire came up from there," she wrote. "I got some action right away on that Cow Creek fire but when the men got up there . . . they couldn't do a thing because all their gas tanks on their equipment had exploded . . . everything was so dry and the wind was so strong."

The next day, virtually all the Company's young men were out on the fire lines. Building a fire of any kind or even striking a match in Hilt was forbidden, since every pumper truck in town was gone. Grandfather had to sneak cigarettes in the bathroom like a guilty eighth grader. For several days, we watched the Beaver Creek drainage burn, glowing red in the night behind the ridgeline, the smoke reducing visibility to fifty yards by day. The fires burned within two miles of Yreka, and on one day alone, between three o'clock in the afternoon and ten o'clock at night, consumed twenty thousand acres of timber.

We sat on the porch, watching the wavering red glow behind Bullion Mountain. Forty years later, one Forest Service firefighter remembered his plane trip from Montana, flying toward the Siskiyous in a DC-6, just after dark. He looked westward at what he thought was a California sunset, only to watch it become a vision of flames on top of the mountains, as far as he could see.

At Hilt, Fruit Growers welcomed the firefighters, put them up for the night, fed them, and the next morning stuffed them into buses and sent them out to the fire lines. Sparks started fires twenty miles ahead of the lines; fallers (chainsaw operators) found themselves sixty miles from where they had started, thirty hours away from their last food, their last sleep.

We scanned the jagged horizon with Grandfather's binoculars, picked out the single light of Hungry Creek Lookout, and knew that the fire could not be too close, or she would already have left. In fact, that first night Clara knew almost as little as we did.

> I couldn't see the Haystack Fire itself, but the smoke went boiling northeast right over my lookout. It looked like a big ostrich plume The next day we had another storm but no rain. Between the smoke of Sterling, Cow Creek, and the Haystack Fires was an opening, and I saw the strike in Dutch Creek (a tributary of

> Beaver Creek) which set the Dutch Creek Fire. There was no one available to go to that fire except a state tanker, which was sitting down at the recreation hall on the Klamath River. By the time they got there, there was nothing they could do to hold it . . . and it burned toward Dry Lake. Then, it reversed itself and came back down the mountain and across Beaver Creek and up over Buckhorn Mountain. We thought the lookout had burned, but the fire divided and went on each side of it . . . we were all sent down from the lookouts the next day because the Dutch Creek Fire was coming right toward us about three miles away.

The Dutch Creek Fire never reached the Hungry Creek Lookout, although the fire outran and burned up bulldozers and dissolved abandoned chainsaws into ash. Finally, ranks of D-9 caterpillars put their blades down and worked their way to the head of the Beaver Creek drainage, to the Siskiyou Divide, and the twenty miles of fire line on the ridges at last turned the flames hack upon themselves.

In October, the Forest Service took stock of its losses and auctioned the first batch of fire-damaged timber. But for Fruit Growers, the great fires had been very serious indeed. Besides the fires that devastated Company lands in Beaver Creek, the great swath of lightning-caused fires sent the Company's Indian Creek holdings, west of Yreka, up in flames. Tens of millions of board feet of timber, that the Company thought would feed the sawmill for fifty years, now stood fire-killed and if not salvaged within a year, two at the very outside, would he lost forever to bugs and rot. That fall, the Company built a logging camp near Dutch Creek, the last logging camp ever built in the county. Late in December, a warm wet storm blew in from the Pacific; rain fell to the top of Mount Ashland. The Flood of '55 took out the lower section of Beaver Creek Road, wiped out the logging camp, and the charred, hydrophobic soils were scoured away down the canyons.

It was a dark winter for everybody—for Aunt Jo, exiled to Sacramento; for Elizabeth and me, who missed her and were bored without her; for Grandmother and Mother, who could neither stop Grandfather's tempers and sulks, nor escape them; and for Grandfather himself, who in the midst of his anger and hurt and self-imposed isolation, failed to recognize that the aftermath of the Haystack Burn would in fact bring the Company within reach of so much timber that Grandfather's job would even have been safe if Jo had danced naked at noontime around the flagpole on Front Street.

CHAPTER 10

> I can call it all back, and make it as real as it ever was, and as blessed. I can call back the prairie, and its loneliness and peace, and a vast hawk hanging motionless in the sky.
>
> MARK TWAIN

The flood of '55 did little damage in Hilt itself. I remember few days of ceaseless warm rain and the felicity of running home from school through a richness of puddles that rose to engulf whole streets. At night, the house creaked under strong winds. Lightning flashed behind the clouds and the thunder was drowned out by the rain pounding on the roof. A few shingles lifted over the woodshed, and Grandmother put a coffee can under the leak. One of the small creeks that ran through Hilt threatened to overflow its culverts, and Company men struggled to keep them clear of debris. My sister and I liked the flood and were sorry that Aunt Jo was missing it.

Jo was oddly content at the Fairhaven Home for Unwed Mothers. In a large Edwardian house on a wide shaded boulevard, she lived a quiet life. She made friends. The home had no sign on its door, of course. The young women who came there to wait out their pregnancies used assumed names—Jo called herself "Martha"—and sometimes were allowed to go shopping in pairs. The place was an accredited hospital, and doctors came to give regular examinations and finally to help with the deliveries.

Jo could not write to her family, lest the return address and the postmark lead to questions. The official story was that she had gone back to school at San Jose State. Telephone calls were too expensive for Jo.

Grandmother could not make a call from the house, for Grandfather always scrutinized the telephone bill minutely, convinced that Siskiyou Telephone was plotting to cheat him. Nor could she telephone from the pay phone on the store porch—someone might casually mention it to Grandfather, or just begin to wonder who Mrs. Roush was calling. So she waited, gnawed with worry.

Grandfather's sister Fanny and her husband, Clare, lived in Sacramento. Because Fanny was already suffering from the dementia that would eventually kill her, Grandmother wrote confidentially to Clare, asking him to please visit Jo and let her know how she was. Clare wrote a scorching letter in reply—to Grandfather. In it, he denounced Grandmother for attempting to implicate him in Jo's folly. Why, he wrote, if he went to visit her, people might think *he* was the father! Clare was short, bald, and seventy and wore his trousers cinched up under his armpits, and the thought of him being taken for Jo's paramour should have been hilarious. But none of us laughed as Grandfather slapped the letter down in front of Grandmother.

She was standing at the drainboard, cutting up vegetables for supper, and now and then slipping Elizabeth or me a piece of carrot. When Grandfather started shouting, we retreated hastily to the kitchen doorway, as Mother hurried past us in a vain attempt to shush him, which only turned his anger on her, too.

"Believe you me, this is MY house, and what I say here GOES, and if you two don't like it, you can just get out!"

"You don't mean that, Dad," Mother said quietly. He stared at her with mad eyes, almost as though he had never seen her before. He turned back to Grandmother.

She was leaning heavily on the big broad-bladed knife she used for chopping. She stared down at it, tears dripping onto the celery and carrots, as Grandfather shouted at her. The words were like waves, washing over us: Won't stand for this! How Goddamn stupid can you get? This is all your fault, anyway! You always spoiled her! I should have mopped up the floor with her a long time ago!

Grandmother never looked up at him, and when he started to repeat himself, she began chopping again, with little rocking motions of the knife blade, macerating the celery.

Supper was grim. Grandfather's onslaught of epithets continued even as he shoveled forkfuls of scalloped potatoes into his mouth. When he finally inhaled at just the wrong time and began to choke, we all froze—Elizabeth and I because that was our habitual posture during Grandfather's frequent mealtime choking bouts. But usually one of the women would rise to slap Grandfather on the back and fetch a glass of water. Now neither of them moved as his complexion darkened from red to purple and his coughing spewed little pieces of potato all over the checkered tablecloth. His streaming eyes lost focus behind his thick glasses. Finally he gave a last gurgling heave and began to breathe again. Mother's fork, which had hung in midair, completed its journey to her mouth, and her lips snapped shut over the small forkful of salad. Her eyes, hard as brown marbles, never left his face. Grandmother rose slowly and went into the kitchen, her lips folded tight. When she returned with a glass of water, her face bore an expression I had never seen on it before—regret, and resignation.

In March, as the snows melted in the mountains, the Company hired more loggers and logging engineers and foresters and began to rebuild the flood-damaged roads and break new ones into the woods. They bought "crummies" to bus the loggers into the woods every morning, and Mrs. Benson lost her job as logging camp cook. In the spring, the telephone rang late one night, startling me out of sleep. The baby was born, Jo said, and she wanted to keep it. The child had been born with red hair, and the husband and wife originally chosen as adoptive parents were no longer suitable. There would be a delay in the adoption while the agency scouted around for another family. Elizabeth and I crept out into the front room, as Grandfather's voice, habitually loud on the telephone,rose to a boom, angry.

"Well, you don't live on Dobie Street!" he yelled at the mouthpiece. "You won't bring it here, that I can tell you! You're not a Goddamn Dago!"

I hung on Mother's pink bathrobe, so hard that she looked down and began to pry my fingers loose. Elizabeth had already fallen back to sleep on the couch, her thumb in her mouth. I lay down beside her, Grandfather's shouts dying to a rumble as I drifted into a shivering sleep.

A more restrained discussion took place the next day. Mother offered to adopt the baby. Grandmother wanted Jo to bring the child home, never mind what the rest of Hilt thought. Everyone was tired, and Grandfather's voice was hoarse. He no longer shouted. He only refused, and refused, and refused. No, he said, never, never. A week later, Grandmother sent me off to school with a note pinned to my sweater which instructed Mrs. Davenport to keep me after school until she came to fetch me.

The day began inauspiciously, as I managed to clog one of the girls' toilets with a great many of those institutional squares of toilet paper. When flushing not only did not remove them, but seemed to encourage the toilet to regurgitate them, I fled back to the classroom, pretending nothing had happened. But Mrs. Davenport soon led me back to the scene of the crime, demanding an explanation, and I could only shrug, and state with perfect candor, "I don't know."

When the final bell rang that afternoon, and the other kids went home, I feared that they must all think that I was being kept after school for clogging the toilet. I wished, profoundly, to be somewhere, anywhere, else. What if Mrs. Davenport told Grandmother about it? Then Grandmother would probably pull a switch from a handy tree and whip me ahead of her, all the way home. Alone, I wandered down the big dark front hall of the school, looking out the west-facing double doors, over the broad porch above the basketball court.

The March day was cold and windy, and the empty old building groaned under the gusts. Out on the playground, I saw Robbie walking with Paula, another second-grader. She made no effort to keep up with him, as I would have. Instead, he was slowing to her pace. The dark curls below her ears trailed gently down onto the soft olive skin of her cheeks, and I knew that I hated her.

I walked back past the lockers reserved for the seventh and eighth graders. The kids who brought their lunches ate in the seventh and eighth graders' room in bad weather, and I had brought my lunch that day. A group of the big boys laughed at me when I dropped my milk thermos. I heard the glass shatter inside, but the teacher on lunch duty didn't even look up as I fetched paper towels from the girls' restroom and tried to mop up the mess. While I was gone, someone ate my oatmeal cookies.

Now, I walked down the darkened halls and back into my own empty classroom. Mrs. Davenport looked up from grading a pile of papers and suddenly smiled at me. "Well," she said, "It's been quite a day, hasn't it?"

A long shaft of afternoon sunlight pierced the overcast sky, flooding down the south hall, and poured into the room, all the way back to the stage, as Grandmother came into the room at last, a sweater over her arm. She took my hand and we walked down the hall and outside, my lunch box with its shattered thermos inside banging against my knee. Grandmother didn't say anything about the toilet.

We walked briskly, while Grandmother stared sternly ahead, her face set and lips compressed, but she didn't stop to pull a switch from a tree, even though we passed several good ones, so I knew that if she was angry, it was not at me. We took the short cuts, through the alleys and across the old playground behind our alley, and went in the back gate. I ran to open the back door, and as it swung open, I felt rather than heard the old piano, its loud pedal pressed down all the way, launch into a deep familiar rhythm, followed by the high, clear melodic line and Jo's full-throated voice ringing out through the whole house with "Heart and Soul."

I swung around the bedroom door and saw my aunt at the piano with Elizabeth beside her, and they made room for me on the old seat, and we sang the rest of the song together, surrounded by music.

In that spring of 1956, I learned to ride my new red bicycle, really ride it, without training wheels. Jo hung onto the rear of the seat, steadying me as I pedaled, until one day I at last found the right combination of speed and balance, so that when Jo finally, invisibly, let go of the seat, I continued to pedal and to sail on past houses and trees, down the sidewalk, on my own, her voice cheering me on.

When Jo came home, she told us about our red-haired girl cousin. I don't believe my mother knew that she had told us; she would have considered it dangerous to entrust such a bombshell to the potentially loose lips of two children. But we never told anyone. Jo said she had given the baby to some people who couldn't have children of their own, some people who needed the baby more than she did. She didn't tell us that she had ridden in a car, carrying her child in her arms, to a run-down section of Sacramento. On the way, Jo had asked the matron from the home what would happen if now, at this moment, she decided that she wanted to keep the baby.

"You aren't going to do that," said the woman, "or you wouldn't be here now."

She had been led up the steps of an old gray house and into a room fixed up like an office. Papers were placed before her and a pen. She signed them, and then another woman—an anonymous, hard-faced middle-aged woman—had entered the room and taken the baby from Jo's arms and carried her out of the room. Jo panicked, and stood up, and started to go after the woman, hut the matron put a hand on her arm, and the driver, a man, stood in front of the door. Jo wondered, at that moment, if indeed the child was going to be adopted, or if the hard-faced woman was even now drowning her child in a sink somewhere in the house. But surely, they wouldn't do that. After all, she had signed those papers. She looked back at the desk, but the papers were gone.

For the first month after Jo returned to Hilt, she was pale and thinner than we had ever seen her. Mother told us not to play too roughly around her, because she had been very ill. We nodded and thought we understood. We had both had chicken pox. We knew what it was to be sick. By April—on the surface at least—Jo's life seemed as busy and lively as it had ever been. She went to the spring session of the business college in Medford. She made friends there. One of them had a cousin in the Air Force, and she introduced him to Jo. Late in June, just as Jo came home for the summer and the Air Force cousin prepared to leave for a tour of duty in Turkey, he and Jo became engaged.

Suddenly, Jo's whole status changed. With the addition of a modest diamond ring to her finger, she was redeemed. From being Jo, the Fallen Woman Now a Burden to Her Family, she became Jo, the Affianced. Even Grandfather unbent enough to acknowledge her existence. He grew, in fact, almost cheerful.

The wedding was planned for May of 1957. She would be married in the Episcopal Church in Yreka, and the reception would be in Hilt, and Grandmother would make the cake and the wedding dress. From out of some closet, the dressmaker's dummy we called Dolly emerged, and Grandmother began work on Jo's trousseau.

With the unswerving selfishness of a child, I would have remembered that summer as a happy one simply because *I* was happy, but contentment seemed to be in the air. Jo was suddenly jolly again—the gay companion of yore. She got the croquet set out of the woodshed and set up a course that ran from the back yard under the cherry tree to the front yard under the big spirea bush. After supper, the front porch again teemed with the teenagers of Hilt, who gathered to listen to a young Latino named Tony play his guitar and sing "Mary Ann," while the wooden balls clacked below. Once more Jo rode away into the dusk with friends, but now she rode with carloads of girls, and her demeanor around the young men with the cigarette packs rolled up in their T-shirt sleeves was one of amused tolerance, a duenna guarding her charges. She moved serenely through the day on a schedule all her own. She unearthed her old bow from the back of her closet and practiced archery in the back yard and mailed photos of herself, bow in hand and captioned "The Huntress," to her fiance in Turkey. She gave us her collection of Superman and Wonder Woman comic books, and we read them lying on her bed while she practiced the piano.

She wore a lot of eyeliner and brushed mascara onto her lashes and devised dance costumes for us that Grandmother sewed. She taught us new dance routines and songs, and we all performed in a big community talent show in the Clubhouse that summer. On weekends, Jo and Mother gave each other permanents and cut and set each other's hair.

Carl Goddard, Jo's intended, was a dairy farmer's son from the little town of Talent in Oregon's Rogue River Valley. Ormy and Edythe had managed their little farm on Wagner Creek for forty years. Both were descended from well-to-do pioneer families—doctors, county assessors, and land owners for three generations. Carl had a bad complexion, a military haircut, and bore a slight resemblance to Elvis Presley. We liked him, and he liked children. The Air Force had trained him as a cook, and Grandfather, who began in this year to take over the kitchen on Sunday mornings, and to turn out bacon and eggs and gigantic hotcakes from a recipe of his own invention, was so relieved to find that his future son-in-law was both white and from a respectable family that he immediately issued a standing invitation to Carl's parents to come for Sunday dinner whenever they wanted. For our part, we began stopping by the farm to visit on our way back from shopping trips to Medford, and Elizabeth and I discovered a wondrous place full of cows, irrigation ditches, and a great barn with a real hayloft.

Ormy and Edythe had grown up attending Sunday meetings at a place called the Universal Mental Liberty Hall, where the only rules for speakers were that the audience could ask questions after the sermon or talk. Other churches considered the Universals a nest of infidels. Grandfather was a little startled to find that this elderly farmer and his wife (for Carl, their only child, had been an unexpected production of their middle age) were what his old Sunday school teacher would have called "dangerous freethinkers." But the political discussions that now raged around the card table in the kitchen, where Grandfather held court and mixed drinks before the roast was served on Sundays, began to fascinate me.

Primed with a couple of drinks, Grandfather on these occasions indulged his new hobby of mixing cocktails. He had invested in a stainless-steel cocktail shaker, shot glasses, long-handled spoons, and martini glasses. He acquired rows of jars and bottles containing cocktail olives and onions, grenadine syrup and creme de menthe and gin and vermouth and vodka and several types of whiskey. He built martinis with panache and presented them to drinkers with a flourish, like Jonas Salk handing the world the vaccine for polio. But the best thing about the cocktail making, as far as I was concerned, was the talk that started so easily under the lubricating influence of those bright liquids in the glasses and bottles. Religion, politics, the hydrogen bomb, the elections to come. I sat on the stool beside the trash burner, out of the way, with my feet tucked up on the high rungs, and listened as the new members of the family dropped bombshells of conversation into the autocracy that had been our household. Opinions that Grandfather would have ridiculed or firmly quashed had they come from Mother or Grandmother or Jo not only were allowed to see the light of day, but were listened to thoughtfully and respectfully.

"Do you know what an infidel is?" Ormy would ask Grandfather over a stiff whisky and soda, in the middle of a conversation about the Arabs and Israelis. "It's a person that don't believe with the prevailing belief. If you were a Christian in China, you'd be an infidel. If it's against the prevailing religion, that's an infidel." I held my breath, but Grandfather did not yell at this iconoclast, or tell him he was stupid. He listened to Ormy and actually *talked* to him, and to Carl, without shouting. Somehow, these outsiders were different from us. It would be years before I realized the real difference—they were men, and independent men, at that: not employees who had to agree with him, not storekeepers who wanted to sell him something and therefore agreed with him even though their eyes revealed their contempt. They were men, and not women, and therefore equals, and therefore people.

Sometimes when we visited the farm, Ormy got out his bagpipes, and as he blew the bag full of air and the penetrating shrill whine filled the room and resolved itself into the opening notes of "Amazing Grace," I felt all the hair on my arms stand up. I wanted to march off and do wonderful things, fight in a great battle, preferably on top of a war-horse. As a soldier in France in the Great War, Ormy had seen battles, but now, he said, he was a pacifist, a word that always made Grandfather bristle.

Carl's upbringing in this small, easy-going family, who actually seemed to like each other, may be all the explanation needed for why Jo wanted to marry him. To watch Carl and Ormy playing cribbage together, bantering like old friends instead of parent and child, was to see something new to all of us in the lives of men. For his part, perhaps Carl looked at Jo and saw only her gaiety and beauty and love of life and believed them more than enough. He knew about the baby, Jo told her sister, and he didn't care.

"He doesn't think things like that matter, if two people love each other," Jo said of her betrothed, as Mother nodded her agreement.

"Although," Jo continued, "I'd marry the Devil himself if he offered to get me out from under Dad's roof—even if we do have to wait a year, until he gets back from Turkey."

In that most remarkable summer, even our mother, of all people, suddenly seemed younger and giddier and altogether more fun. She took evening lessons in modern jazz dance from Karen Adele. She choreographed a duet with Jo, although they only performed it before the family. She went out more often to the Company dances and, in a sudden access of family togetherness, everyone in the house went to the Fireman's Ball that year, in the Warrens Building. Elizabeth and I were the only children present, but we danced out on the floor with the grown-ups and finally fell asleep on a pile of coats in a corner, until we were taken up and carried home at two in the morning.

On weekday mornings, Mother crossed the railroad tracks to the Company's spanking new office building almost eagerly, we thought. Beyond our sight, the young foresters thundered down into the new basement offices, early in the morning, then headed out into the woods in the Company's dark green pickup trucks and were seen no more all day. But Mother knew where their desks were, and she knew their names, and she saw them just often enough, that summer, to learn their faces. By the winter, when the woods department came back inside to draw maps and design roads and manipulate the timber cruise data they gathered all summer, Mother found more and more excuses to go down those stairs and sharpen ten or twenty pencils on the pencil sharpener that just happened to be mounted on the wall across from the desk of a tall twenty-seven-year-old logging engineer, with blue eyes and a cleft in his strong chin. At the age of thirty-three, Mother—the prim, the neat, the businesslike—was determinedly flirting with a man six years younger than herself.

Grandfather lived among us but not of us. He read novels by Lloyd C. Douglas and Thomas B. Costain deep into the night. He came home from work and hid behind the newspapers, ignoring the laughter and guitar music on the front porch. He went fishing by himself on winter days, when the steelhead were running in the river, and never even asked Grandmother if she wanted to come along. Grandmother relaxed visibly after he had gone, writing letters, putting her feet up and reading the newspapers, sewing doll clothes for us from patterns of her own making.

Grandfather gobbled his supper and left the table without comment. Occasionally he flew into a rage at the perfidy of politicians, as he watched the evening newscast on television or read the editorials in the *Sacramento Bee,* "that Commie rag." He hated Nixon, however, which I found confusing, so that by the time I walked up to the Clubhouse with him on the first Tuesday in November to watch him cast his vote after work, I was convinced that Eisenhower was running *against* Nixon.

More and more, Grandfather took Elizabeth and me with him in those days—when he voted, when he walked down to the box factory on weekends, when he drove up to Gino's State Line Service Station and Liquor Store to spend an hour consulting with Gino Trinca on the best buys in gin and vermouth. While Gino placed the bottles carefully inside brown paper bags, twisting the paper around their necks, Grandfather leaned against the bar and sipped a beer and bragged about us, his grandchildren, to Gino and anybody else nursing a beer in the bar on a long afternoon. We, he told Gino, would be the first Roushes to graduate from college. Louise, there, she wants to be a scientist—she knows all the planets, already. And she can read anything you want to show her and knows all the names of those dinosaurs and how far it is to Mars.

While Grandfather talked, Elizabeth and I climbed up on the barstools and sipped Shirley Temples and stared up at the wall behind the bar. Carefully tacked to the knotty pine woodwork was a remarkable collection of pinup calendars, collected, apparently, from every liquor salesman who ever stepped off the highway. The unclothed women fascinated me, with their lush pink bodies, their huge breasts, their flat bellies and perfect buttocks. Their smiling, lipsticked faces and perfectly coifed hair were like those in the Breck Shampoo ads, but their bodies were nothing at all like a shampoo ad.

In September, I entered a new school, built just below the old one. Brick and steel and glass, its unimaginable size and luxury awed us. With its furnace room and a system of pipes circulating hot water beneath the linoleum tile floors, we could run about comfortably in stocking feet on winter days. Every classroom had a big sink and a drinking fountain, low for the first graders, much higher for the seventh and eighth graders. Long, wide counters and cupboards and bookcases ran around the room, and the huge windows let in the light. Drapes could be drawn to darken the room for movies or slide shows. Banks of recessed fluorescent lights covered the entire ceiling in white light. Acoustical tiles lined the upper walls. There were long open closets for coats and boots, locking closets for the teachers, a big bin for balls and bats. There was a lunch room, where long tables and benches with Formica tops folded down from one wall, and a big kitchen, where Mrs. Benson now cooked hot lunches for children instead of loggers.

In the long hallway, signs informed us that the school was an official Fallout Shelter. In case of attack, we were to file into the hallway and sit on the floor, while the teachers pulled the drapes and shut the doors behind us. We had fire drills in which the earsplitting alarm horn blared and we marched outside onto the asphalted playground to stand shivering and giggling in the morning chill, before the cheerful all-clear ring of the electric bells.

A plaque on the new building informed us that this was the Stella King Elementary School. Mrs. King, the retired long-time principal at Hilt, had been greatly beloved, especially by the Italian immigrants, whose children she treated with such patience and kindness that at least one child was named for her. A staunch supporter of Franklin Roosevelt, proud to call herself a New Dealer, she was a plain, shy widow whose heroines were Eleanor Roosevelt and Frances Perkins. She raised two children by herself and trembled every time she had to give a graduation speech. "I can write a good speech," she told Grandmother once, "but I'm terrified to deliver it."

Now, her name stood out in large steel letters on the east wall of the school, right behind the flagpole, a lasting affront to Grandfather, who loathed her for her politics.

There were new faces in our second grade class—kids from Westwood who swelled our ranks so that the school hired Mrs. Wesner from Hornbrook to teach us, even though she lacked several credits of what the state of California considered necessary to teach seven-year-olds.

Mrs. Wesner had short dark hair and a mustache and brought two parakeets in a cage, who rode back and forth with her each Monday and Friday. During the week, they lived in a cage on the counter at the back of the classroom. Children who finished their work satisfactorily were allowed to feed Johnny and Janey, or take one of them out of the cage and back to their desks, where the birds amused themselves by flipping pencils out of the pencil groove and chasing them down the slope.

Of the second grade, I remember Carleen DeClerck, who skipped into our class from first grade and promptly fell off a fence and broke her right arm. She amazed us by writing ambidextrously from then on. I remember Ronnie Chase, who fell out of a tree and knocked one of his big front teeth out. The enormous silver replacement tooth made him an instant celebrity. I remember the disappearance of Janey, who was eaten by Mrs. Wesner's cat one weekend, and Mother crankily attempting to put together the materials required for a class art project—a coffee can, raffia, and a circle cut from an old inner tube—one evening after work. What could you expect, she muttered, from someone who wasn't even A Full-Fledged Teacher and kept Filthy Birds That Probably Had Parrot Fever in the classroom.

I remember tadpoles and caddis fly larvae that existed in jars in our classroom windows, until they were forgotten over the weekend and died and were unceremoniously poured out into the weeds below, on Monday. I remember recesses when we hauled large round rocks up the hill to a juniper tree near the swings and piled them around a depression at its base, eventually building a wall two feet high. On cold windy days we crawled down into the sheltered hole to stay warm until the bell rang. The teachers were afraid it would fall on us and made us take it apart again.

Then and in later grades I remember an endless series of shots—polio shots, and diphtheria shots, and tetanus shots, and smallpox inoculations, which, as they began to "take," turned into grotesque, pus-filled scabs, which we eagerly compared, peeling down a corner of the gauze bandages on our arms.

I remember playing on the monkey bars and learning to hang by my knees and swing across on the horizontal ladder, bar by slippery bar, and then rubbing dirt into my blisters to toughen my hands. There were dizzy whirls on the merry-go-round and sitting on pieces of waxed paper from our lunch boxes as we slid down the big slide. There was jumping rope and learning all the old skipping rhymes that our mothers had learned from their schoolmates.

Charlie Chaplin went to France
To teach the ladies how to hula-hula dance.
First on their heels, then on their toes,

Around and around and around she goes.
Salute to the captain, bow to the Queen,
Turn your back on the dirty submarine!

We played jacks and hopscotch and foursquare and tetherball. There was plenty of room for us all, and we could avoid the eighth graders now if we wanted to. But sometimes, the big kids even let us play workups with them.

And we looked out the tall windows, and watched the seasons change, and the gray November skies gradually lowering over Skunk's Peak, that would bring the unsullied snow to Watertank Hill. I remember December days when we stared out at the blinding whiteness until sparks flew behind our eyes. I remember the spring, when we craned our necks to take in the whole of the windy sky and the black turkey vultures pasted on it, growing smaller and smaller as the sun warmed their wings.

We came home after school to a house smelling of cookies, and we grabbed handfuls of them and went out to ride our bikes on the sidewalks and down the alleys, in search of adventures. The old roses bloomed on the front fences, and the robins sang songs of territory and lust in the trees, and I felt, as I flung my arms out and rode no-hands down the hills, sudden surges of mindless, unreasoning joy.

After supper, we went out again and rode through the neighborhoods until dusk, as fast as I could make my little bike go, leaving Elizabeth far behind on her tricycle. Sometimes I looked out at a pile of thunderheads behind Hungry Creek Lookout, shaped like dragons or horses or gargoyles, all glowing red and gold with the sun behind them, and my seven-year-old brain came alive for the first time with the discovery that I was *me*, myself, and no one else, and that I was happy in this place that was all the world there was, among the silent brown hills, beneath the roaring branches of the old trees along Front Street.

1. In 1972 the mill complex at Hilt was still working, and I took this snapshot from the sidewalk in front of my grandmother's old house.

2. Part of Hilt townsite in 2000. Mount Ashland is on the horizon.

3. Hilt townsite in 2000, looking west from Front Street across the railroad tracks.

4. Louise with first steelhead, May 1961.

5. Grandmother with steelhead, about 1955.

6. Barbara, Martha, and Bill Roush, near Susanville, California, 1924.

7. Birthday party under the white poplar. left to right: Stevie Flyte, Robbie Flyte, Caroline Ruger, Elizabeth, Louise.

8. Left, John Brannon; right, Charlie Brannon, as young boys, near Lakeside, California, in the 1930s. Pappy is the second man from the left.

9. Barbara with Jo, late 1930s.

10. Center, John Brannon as a young man in the logging woods near Wren, Oregon.

11. Elizabeth and Louise at a ballet class in the old Clubhouse at Hilt, 1956.

12. Aunt Jo and Mother, 1958.

13. Mother and Daddy, about 1960.

14. Liz and Tommy under the black locust tree, summer 1962.

15. Louise at the gate of the house on Front Street, about 1952.

16. Grandfather in full fishing regalia, 1954.

Part 2: Black Locust

CHAPTER 11

When asked why he did not become a father, Thales answered, "Because I am fond of children."
DIOGENES LAERTIUS

John Brannon, the handsome young forest engineer who had caught Mother's eye, did not ask her out on a date until February of 1957. In spite of all the eye contact over the pencil sharpener, Mother kept her expectations low as she sat beside him at last in Yreka's Broadway Theater. She told herself firmly that what he probably wanted was her influence with Barbara Alphonse, a younger, prettier, childless, *eligible* woman. He inadvertently encouraged this suspicion by asking questions about the Alphonse sisters, and as she said goodnight to him at the front door, she was utterly convinced that she would never see him again outside the office.

John had graduated from Oregon State University with a degree in logging engineering in 1951. In college, and for several years afterward, he and his father and brother were independent loggers—gyppos—in the rain forests around the Willamette Valley. His father had been a farmer and well-driller from east of San Diego, California, who moved to Oregon in the midst of World War II.

John and his only brother Charlie—for many years called Bo after John's first lisping pronunciation of "brother"—started logging while they were still in their teens, and it was John—lugging home his college texts on logging—who masterminded the family enterprise. Charlie drove the log truck, and Pappy, as they called their father, operated the Cat and the loader and maintained all the machinery.

All this we eventually learned because John did ask Mother out again. And again. And one day, Mother brought John home for lunch, and we got our first good look at him. Mother's beau stood an inch and a half over six feet. He was the tallest human being we had ever seen, and of course he towered over Mother. His eyes were a light, intense blue, set under heavy brow ridges and a high forehead, which looked even higher thanks to the early receding of his fine, light brown hair. Slender, with rather narrow shoulders, he had big, tanned hands with strong wrists, and his fingers were long and well shaped, calloused from a lifetime of outdoor work. His hips were narrow and his legs were long, long, long. He had an infectious laugh, and his large white incisors protruded just the slightest bit, pushing against his too-short upper lip. His grin came easily. Elizabeth and I fell in love with him at once.

There is no more romantic creature in the world than a seven-year-old girl. I didn't know the first thing about sex, but I knew romance when I saw it, and that Mother was nuts about John, and he about her, was obvious to me at that first lunch, as they blushed over the soup and tried not to look at each other. Elizabeth and I exchanged sophisticated glances, and Jo raised her curved right eyebrow at us.

When Carl returned from Turkey in April, to prepare for his wedding to Jo, we saw the difference immediately. Jo and Carl were like old friends, like teenagers who agree to go steady so they can be sure of a date on Saturday night. Mother and John were In Love and looked it. Elizabeth and I thought it was the cutest thing we had ever seen. By May, we were calling him Daddy.

Grandfather at first looked narrowly at John over the newspaper and made discreet inquiries about his antecedents, habits, and income. John did not disappoint under the scrutiny. He didn't smoke—unusual then—nor dip snoose or chew (the most common vices of loggers). He lived quietly and didn't frequent bars or chase women. He was young, but already making a good salary. He passed.

John was different from anyone we had ever known. He didn't wear a hat when he came calling. He didn't wear a fedora in town, at a time when most men did. When he came to take Mother out, he wore good slacks with a belt, but no tie or coat. Instead, he wore a Western shirt with snap buttons, and a bolo tie, the first we had ever seen. He began taking all three of us on weekend outings—to the museum in Yreka, to the matinee. Sometimes we stopped by his apartment in Yreka so he could pick up something or offer us a drink of water.

The first time Elizabeth and I saw the little bachelor apartment, we wanted to take him home with us permanently. We agreed that he shouldn't have to spend another lonely night, another insanely boring weekend, in this impersonal bed-sitter, up a dim staircase from the back of an old Victorian house in Yreka. Furnished with the landlord's castoffs and the few pieces of furniture he had been able to bring down from Oregon in his red Chevy pickup truck, the dark little place echoed unbearably. The dull walls were bare, the old lace curtains yellowing.

In the bedroom, a brown portable electric radio sat on the nightstand. There was no television; and what was home without the cozy brightness and noise of a TV? What was home without a warmly lit kitchen and a houseful of people? We felt sorry for him.

One Sunday, John took us all to a movie in Yreka, and right in the middle of the show, just as Robert Mitchum was about to kiss Deborah Kerr, even though she was a nun, Elizabeth turned around in her seat and looked up at John's face in the dark and said, in a voice as loud and startling as a peacock's, "Daddy, when are you and Mommy going to get married?" As Mother said "Shhh!" and scrunched down in her seat with embarrassment, Elizabeth turned around and sat back down with a thump onto the cracked brown upholstery. "There!" She whispered to me, with the air of someone who has done a hard but necessary thing. When we emerged from the theater, Mother's face and John's were still as red as the usher's flashlight beam.

In May of 1957, Jo and Carl were married at St. Mark's in Yreka. I remember a packed church and Jo in a long white dress with a train and a net veil. Elizabeth and I were flower girls, carrying small baskets in gloved hands, wearing flowered hats. Our dresses were buoyed by crinolines, our shoes were white patent-leather Mary Janes. Mother was matron of honor. Back in Hilt for the reception, Jo and Carl—handsome in his Air Force uniform—cut into the three-tiered cake, while Grandmother took pictures with her new color camera. Father

Robinson drank martinis in his clerical collar, while Elizabeth and I ran in and out of the house with our three girl cousins from Mount Shasta. Jo fed her new husband slices of cake and opened piles of presents and looked happy. But when it was all over and Jo had thrown the bouquet and tripped down the front walk in her new traveling suit to the curb, where Carl waited to hand her into his car, it hit me. Our aunt was leaving us. We would see her when she came back from her honeymoon to the Oregon coast, but she would not stay. She would load her wedding presents into that car, and she and her new husband would go away to Oklahoma City, where Carl had orders to report to the nearby Air Force base. I ran down the walk after her and hugged her through the open car window and began to cry as the car pulled away.

By suppertime, all the guests had left except John, and when Mother announced, after supper, that John had asked her to marry him, Elizabeth grinned like an idiot, perhaps feeling that it was her doing. I, who had been so pleased that Mother had a boyfriend, was now not so sure about the whole thing. When grown-ups married, I had just realized, Things Changed.

Our mother and John Brannon were married in Yreka that June by a nervous young justice of the peace who kept dropping the prayer book. Mother wore a yellow tweed suit and a matching hat. They came back to Hilt to eat one last lunch, and Daddy sat eating tuna fish in a blinding white shirt and a real tie and thin blue suspenders. He really did own a suit, after all. Then they both climbed into the red pickup and left for a weeklong honeymoon in Oregon. Grandmother watched them go, a half-worried, half-thankful look on her face. Elizabeth and I waved them out of sight.

When Mother and Daddy came back, she took us up to inspect the house we would live in, now that we were a new family. The new house stood two doors up from the Community Church, on the highest point of land on Hilt's south side. On the long sloping lawn grew a black locust tree with a trunk two feet in diameter. Great numbers of these trees grew in the county, another gift of white settlers. Useful as fence posts, resistant to rot, and quick to sprout from root or stump, they bore long, pinnately divided leaves and in spring hummed with bees feeding on the drooping clusters of white flowers. In summer, dark seedpods appeared among the leaves, full of tiny, black, bean-shaped seeds, which were poisonous (something no one thought to mention to us). The branches rose straight up, from fairly high on the trunk, which made them a hard tree to climb, and the smaller branches were armed with long wicked thorns.

All the houses in Hilt had been built before bulldozers came into general use, so houses conformed themselves to the slope of the ground. Our new house faced east, toward Sheldon Rock, and leaned far out over a rocky hillside, so that the front porch stood high off the ground, over a climbing rose bush. Underneath the front porch, a door opened into a tall storage space. While Mother knocked on the door and went in, we inspected the outside.

Between the new house and the church was a house with a remarkable yard, separated from ours by a board fence of pecky incense cedar wood. The front gate was of smooth boards, painted with a garish depiction of the legendary thunderbird, and overarched by a rose trellis. Beyond was a warren of flower beds blazing with dahlias, zinnias, hollyhocks, delphiniums, and snapdragons, each bed surrounded by walls of river-washed rocks—quartz and jasper and serpentine and quartz streaked with iron stains. Here and there, Indian relics—stone mortars and pestles, mostly—surmounted the rock walls, or carried burdens of soil from which nasturtiums cascaded. Between the flowerbeds, tiny patches of lawn flourished. And behind all the flowerbeds, next to the alley, a woman was hanging out laundry on a clothesline. She looked at us.

"You must be Barbara's kids," she said huskily from around a cigarette. We nodded, and came closer. "I'm Ruthie Johnson," she informed us and stuck out a freckled hand. We stared at it, then reached out in turn and tentatively shook it. Ruthie's hair flamed even brighter than Mrs. Sife's. She wore a brightly flowered muumuu and rubber thongs. Her bare legs and neck and upper chest were tanned and thickly freckled, and her face was leathery, with green eyes set deep in crinkled folds. She continued to hang clothes until her basket was empty, then picked it up and started for the house. "I heard tell that some Indians came through here last night," she said, one eye squeezing shut in a wink. "I think if you look in the alley behind the house, you might find some arrowheads." She gestured with her double chin at the alley behind her woodshed.

We ran out the back gate and down the hill to the back of Ruthie's house. Scattered among the dry puddles and crushed rock of the alley were shards of shiny black obsidian. Some of them were indeed arrowheads—not perfect ones, to be sure, but close enough, even if they did have points with broken tips, or were split in half, or just chips. We picked up all we could carry and ran inside the new house, to show them to Mother.

Inside the kitchen, Mother was talking to Marge Chase, who was up on a ladder scrubbing down the walls. We showed off our finds, and Marge looked down at them. "Oh, yeah," she said. '"Ruth and Greg go out to Nevada every chance they get and camp and look for Indian stuff. She keeps all the good ones and tosses the broken ones out in the alley. The kids love it. She's the Cub Scout den mother, you know. She has them doing all kinds of Indian stuff."

We skipped back down to Front Street ahead of Mother. It seemed very strange for her to be home in the middle of the week. She spent most of her time on the telephone, talking to furniture and appliance dealers, when she wasn't ironing Daddy's shirts. She had, of course, quit her job the day she was married. Now she and Daddy shared her narrow bed, jammed up against the wall of our bedroom. Either Elizabeth and I were sound sleepers, or the couple had already learned the trick of silent sex, for we heard no unusual noises in the night except Daddy's rather gentle snores. But the first time Daddy heard the 2:30 a.m. train coming through, he leapt up, his big feet hitting the floor. "WHAT THE HELL IS *THAT?*" his voice demanded.

We grinned into the dark. "Just the train," we said smugly, natives to his tourist.

We showed Daddy our broken arrowheads, and he nodded and said they looked like the Great Basin style, and from his old metal steamer trunk that now sat in our bedroom, he pulled out a shoebox full of pieces of leather and the tips of deer antlers and small flat pieces of obsidian. He picked up a piece of the black glass and, sitting on the front porch steps, proceeded to create a perfect, delicate arrowhead, and then another. He gave them to us, and we tucked them carefully into the new jewelry boxes that one of Daddy's aunts had sent to us.

Daddy was the first flint knapper we ever knew. He had been "chipping arrowheads" as he called his hobby, since he was in high school and without any instruction, he had worked out a method of pressure-flaking obsidian using deer antler tines and a pad of leather to shield his hands. When he read Theodora Kroeber's book *Ishi in Two Worlds* in 1960, he realized that he had reinvented Ishi's exact method of making points and that Ishi, too, had used discarded glass to make arrowheads.

When several local juvenile delinquents set off some sticks of dynamite at the town dump one night, sending the whole town into the streets, Daddy turned around to see Elizabeth and me standing on the splintery sidewalk, in our nightgowns and bare feet.

"Get back in the house," he said sharply. We were startled. He had yelled at us, and yet he had not. Where Grandfather would have roared, he had simply spoken tensely, a couple of decibels above his normal speaking voice, and combined with a narrowing and hardening of his bright blue eyes, it was enough. We obeyed, then leaned out the window of the bedroom, watching and listening.

Daddy had reprimanded us only once before, and that was before he and Mother were married. Stripped down to our underwear in imitation of figure skaters, we had been gliding back and forth across the front room linoleum in our stocking feet, building up speed as our socks buffed the waxed surface, until finally we could slide completely across from the door of the bedroom into the corner by the front door, our heads thrown back and our arms outstretched in ballet positions. Daddy looked up from his seat on the couch and his eyes widened. "Stop that," he said curtly. "Go put your clothes on."

We looked at Grandfather, but he was smoking behind the sport's section, and hadn't even noticed us. Abashed, we retreated, and in our embarrassment climbed out the window to play in the yard, rather than pass Daddy again. Some people just didn't appreciate a good performance when they saw it.

With the walls of the new house washed down and the Chase family moved out, Mother and Daddy filled the house with new furniture and appliances: a pink washer and dryer and a green electric stove. Daddy's old upholstered rocking chair, which had belonged to his grandfather, went in across from the new couch and end tables. Elizabeth and I had our own bedroom, with our same beds, but into the master bedroom next to the kitchen went a new double bed with a bookshelf headboard and two new bureaus, one with a big mirror.

New lamps, a new kitchen table and chairs, watercolors for the front room walls. A maple rocking chair for Mother. No television. Into the laundry room beside the new washer and dryer went a chest freezer, one of the first in town, which Mother filled with a quarter of beef, ordered from the locker plant in Yreka.

Because Daddy's red pickup was too small for all of us, he and Mother went car shopping and paid cash for a red and white 1957 Ford four-door sedan. Then they asked the astonished dealer to take the inside rear door handle off. Daddy was not having any kids playing with the door handles and falling out onto the highway. Elizabeth and I swiftly learned to roll down the windows and reach for the outside latches, but adults who rode in the car complained of feeling like sardines in a can.

All that summer, wedding presents arrived in the mail. Most of them included something for "the girls." We were suddenly celebrities to people we had never met—to Uncle Hank and Aunt Mae, to Cousin Bobby and Aunt Matt and her daughter Cousin Lillie. We heard about people named Grandma Rosa, Aunt Bessie, Cousin Olds, and the Laws cousins. A bewildering assortment of waffle irons, fish forks, embroidered napkins, and figurines piled up. Someone always referred to as AUHnt Helen sent a huge and hideous cut glass VAHz. A woman who signed her name "Tudy" began writing long letters to Mother in a hand so hard to read that Daddy had to translate them. They were from his mother.

Daddy was the older of the two sons of William and Amelia ("Al and Tudy") Brannon. He and his brother Charlie were born near Lakeside in San Diego County. Al had a talent for mechanics; Tudy's maiden name had been Wolff; she had grown up in a nearby German immigrant farm colony, Olivenheim.

We did not meet any of these interesting people until Thanksgiving of 1957. By that time—there was no getting around it—our lives had changed. There were all the new rules, for one thing. Elizabeth and I had to sit up straight at the table and clean our plates, even if we didn't like what was on them. We had to ask to be excused from the table, and permission was not always granted. We had to eat bacon with a *fork*. We had to go to bed at 8:30, long before the other kids were indoors in the summer. There was no television, unless we were visiting Grandmother and Grandfather. And every day we rode past what we now called "the other house," and could scarcely believe that it was no longer ours.

Mother seemed deliriously happy with the situation, and she seemed not only to approve of Daddy's new rules, she even invented a few of her own, and Daddy backed her up. This was new. Elizabeth and I had never felt obliged to mind our mother. Now, we faced what were in effect two new parents, a united front, which took some getting used to, as did Mother's cooking and Daddy's ideas about good food.

Grandmother had never insisted that we clean our plates; she wasn't going to cater to our fussiness, but it was no tragedy to her if we didn't eat our string beans. Now, in addition to Mother's delight in experimenting with her new cooking implements and a shelf full of cookbooks, there was also Daddy to contend with; Daddy, who believed that children should clean their plates, no matter how gruesome its contents. So Mother fixed fried eggplant, because Daddy liked it, and the horrid stuff with its gritty little seeds clung to the roof of my mouth and made it burn. Grandfather liked his bacon so crisp that it shattered if touched by a fork; Daddy liked limp bacon, with disgusting fatty places. He liked his eggs over easy or basted; Elizabeth and I were repulsed by any egg not scrambled or hard boiled, but if Daddy ate fried eggs, we *all* ate fried eggs, whether we liked it or not. Mother no longer consulted our palates.

Everything we had ever done in our short lives was suddenly All Wrong. We had been treated leniently, on the whole, never forced to do anything. We obeyed Miss Adele because she held the secrets of ballet, and we wanted to dance well and to have her approval. And Daddy, too, was a repository of knowledge of esoteric, fascinating subjects entirely new to us. Just following him around every day brought surprises; and as his parents shipped box after box of his old possessions down to Hilt, we realized we were dealing with someone different from anyone we had ever known.

We missed Front Street, although it had never been quite the same to me since Robbie's family moved to Oregon. But the new neighborhood had its good points.

There was Ruthie, of course, who kept us amused by showing off her latest gatherings from pot-hunting expeditions to Nevada. She told us that she hunted for artifacts carrying a walking stick and wearing nothing but a large straw hat and a revolver in a holster, "for rattlesnakes." The idea was startling, to say the least. She showed us bits of basketry and even a couple of scalps she said they had found in a cave.

Just as eccentric in another way was Ruthie's mother, known to all the neighborhood as Nana. She was old and thin and smoked cigarettes in a long holder. She wore "town" dresses every day and a string of pearls around her neck, though she rarely went anywhere. She lived in a hack bedroom with a dresser full of framed photographs, a parakeet in a cage, and a pile of library books on the table beside her bed. In good weather she came out to the lawn, leaning on a cane, and sat in a chaise longue, where she smoked and read and talked to us over the fence. Neighborhood children who met with her approval were allowed to call on her in the afternoons, after her nap, take a Tootsie-Roll Pop from the bowl on her dresser, and converse. Highly favored children were allowed to take Petey the parakeet out of his cage and let him perch on their shoulders.

Nana, whose full name was Geraldine Shelmadine, had led a life as exotic to us as that of an Arab sheik. She had lived in *New York City,* for one thing, and had been married three times and widowed three times. She told us about the things she had seen in New York City—about ticker tape parades and Broadway plays. As she inhaled the smoke from her cigarette, the hollows above her collarbones deepened into valleys. Sometimes she placed a small pill on her tongue. "Nitroglycerin," she told us, "it goes down to your heart and makes a little explosion there." Mother told us it did not, either, but that it meant Nana had heart trouble, and we shouldn't stay too long and make her tired.

Nana was not shy about shooing us out when she had had enough of us, but I always went away feeling better than when I arrived. Perhaps it was because Nana, while never forgetting that we were children, nevertheless spoke to us as though we were reasonable creatures. To me she talked about the books she read, asked me in turn what I was reading, and made suggestions about books she thought I would like. She urged me to read Mark Twain and Louisa May Alcott, and while I lay on my stomach on the lawn howling with laughter over *A Connecticut Yankee,* I was not only enjoying the book, but enjoying the prospect of going to visit Nana the next day and telling her all about it.

For our first Thanksgiving with Daddy, we drove up to Oregon in the new Ford sedan, a daylong journey through the rain. Up through the Umpqua Valley, then over a little pass and down into the Willamette drainage and into yet more rain. At Corvallis, we turned west and wound through a series of wooded glens, through firs so green they looked black, to a tiny post office called Wren. Here, pastures left unmowed for five years became seas of young Douglas-firs. Daddy's parents lived on 160 acres of wooded hills and sheep pasture just north of Wren. We drove up a steep hill in front of an old barn to the house, set on a knoll in a stand of lichen laden white oaks, overlooking the velvet pastures.

We stepped out of the car into the embraces of Tudy, who cooed over Mother and said wasn't she the cutest thing you've ever seen? And how much *shorter* Mother was than Daddy!

Children expect everyone's grandmother to look like their own, but Tudy did not look like anyone's grandmother. She was very tall, only two inches shorter than Daddy. She had very long legs and a very short waist, and she was wearing a pair of slacks and a tailored blouse. Her short dark hair was barely turning gray, and she wore wide bracelets of turquoise and silver on each wrist. She had big horsey false teeth and a more extreme version of Daddy's overbite. She scared the hell out of me.

Fortunately, Pappy was less alarming. He rocked back and forth on moccasined feet, then swept off his porkpie hat and grinned and shook Mother's hand before grabbing Daddy by the arm. "Hello, son," he grinned. His smile was infectious. He laid his hands on top of my head and Elizabeth's. "Hello, gals," he said, and we grinned back. The top of his head was bald, and he was an inch shorter than Tudy. His bifocals were balanced midway down his nose. I liked him at once, but suspected that the last thing in which he was interested was a new pair of grandchildren.

A pair of black and white border collies rose from a rug by the back door and came daintily up to greet us. Daddy introduced them—

Dinah and Dolly. We were ushered through the back door of the house and looked through the kitchen into a living room with a hardwood floor roughly the size of the Serengeti. Mother took off her coat, and as Tudy noticed the gold crucifix hanging around Mother's neck, her large horsey smile froze. Carrying an armload of coats, she beckoned Daddy into a room halfway down a long hall. Was Barbara a Catholic? she demanded. No, she's an Episcopalian, Daddy answered impatiently. Different church altogether. Tudy came back down the hall with her plastered smile intact, but kept staring at Mother's neck as if expecting the Pope to leap out of it.

Months later, Mother explained. The Brannons were extremely suspicious of all proselytizing religions. Daddy blamed this on Grandma Rosa, his paternal grandmother, who in middle age had converted to Christian Science and begun an unrelenting campaign to bring the light of Mary Baker Eddy to her two grandsons. Although by training incapable of being rude to their grandmother, the boys developed a mutual protection strategy. When they saw Rosa approaching, one of them ran off, only to return five minutes later just as Grandma Rosa was getting wound up. "C'mon," the returnee would say, "Father needs us to come help him."

We met Grandma Rosa on this trip; the only time I ever saw her a shapeless white-haired widow who lived in a trailer park in Corvallis. In the few remaining years of her life, she regularly mailed Christian Science children's tracts to Elizabeth and me. Daddy rolled his eyes, but let us keep them. They seemed harmless enough, albeit somewhat silly; we were not tempted by the medical theories or the theology. Daddy disliked all organized religions, especially hierarchical ones; for his part, Pappy liked to relate the story of how, when he was hired to drill a well for a San Diego convent, the Mother Superior tossed a silver religious medal into the hole. When she left, they pulled up the drill and sifted the dirt until they found it.

"Just imagine," he said. "There was people starving not a mile from that convent, and she's throwing money down a hole."

Tudy seemed like a perfect candidate for some of Grandma Rosa's prayers, for her upper spine was curved, and she suffered asthma attacks that sometimes landed her in the hospital. For the first time I understood Daddy's mania about making Elizabeth and me sit up straight at the table. Tudy never went anywhere without her inhaler clutched in her hand or wrapped in a handkerchief in her slacks pocket or in the pocket of the apron she wore in the kitchen. I heard Daddy tell Mother that by now Tudy was probably addicted to the medicines. Certainly, she could not tell an anecdote without several pulls on the inhaler.

The Brannons were a family of storytellers, and Mother, as a new audience, was getting an earful. Al and Tudy told stories about exotic personages named Bud and Woody who lived in places called Blodgett and Kings Valley. Because Al and Tudy owned a boat and went fishing on the coast, we heard stories about clamming and crabbing and nearly being run over by freighters. Tudy told many gory stories about friends and relatives who underwent serious operations or were in labor for forty-eight hours. We found these interesting, although her habit of smacking her dry lips over her long yellow teeth as she told them revolted us.

When the Brannons first came to Wren, Pappy farmed and ran a dairy; then they began to raise sheep—at one time they had fifteen hundred Romneys, well suited to Oregon's wet weather. Then came the gyppo logging. Now, with Daddy working for Fruit Growers and Charlie trying to make a go of it as an independent log hauler, Pappy still had a couple of hundred sheep, but he worked days as a millwright at Woody's mill down by Philomath.

We heard a lot about Daddy's brother Charlie, and when he and his wife Dorothy arrived from Corvallis on Thanksgiving Day, we were thunderstruck. At six feet four inches and 250 pounds, Charlie had Daddy's bright blue eyes and cleft chin, but set on a massive head that merged into a bull neck atop a set of shoulders that filled doorways. He lifted Elizabeth and me up, squealing, one on each wide hand, until our heads bumped the ceiling. Dorothy, six feet tall and eight months pregnant, had brown hair cut very short, and talked about golf and bowling and fishing.

Their three-year-old son Michael was almost as tall as Elizabeth and weighed more. Tudy went to the hall closet and pulled out Lincoln Logs and Tinkertoys and a toy farm with dozens of plastic animals, and we sat on the big round braided rug in that ocean of polished floor and played with him. The grown-ups sat around us and told more stories, now including words like "high-lead" and "Cat" and "hooktender" and mention of places like Sweet Home and Coos Bay and Waldport.

Charlie and Dorothy stayed overnight, taking over the room where Elizabeth and I had slept the night before, so we children all slept in sleeping bags on top of air mattresses in the living room. I can still remember the peculiar smell of Elizabeth's olive-drab Korean War-vintage mummy bag and the whiff of mothballs in mine, a dark blue one with a plaid liner. We had never slept in sleeping bags before, and we loved the adventure. It took the edge off the disappointment we had felt about Thanksgiving dinner.

The dinner consisted of wild ducks stuffed with a wild rice dressing and, for dessert, mincemeat pie made with real venison mincemeat. Today, I would find it a culinary adventure, a nice change. Then, it was a disappointing shock. Elizabeth and I looked at one another and with the innate conservatism of children, felt cheated. Thanksgiving was Grandmother's turkey and dressing, her pumpkin pie and rolls. This was not Thanksgiving. Mother seemed a bit tight-lipped, as well, but that, she said, was because she had seen Tudy washing the dog dishes in the kitchen sink with the people dishes.

The next day, Daddy took us on a walk around the place. We walked uphill behind the house through groves of dripping, bare oak trees. We found the sheep, and as they moved away from us, we noticed an old ewe, lying dead in a hillside ditch. "Wait here," Daddy told us, and went over to look at her. It happened like that sometimes, he told us when he came back, when sheep lay down in a low place and the rain weighed down their heavy wool. If they lay down facing uphill, they sometimes couldn't get up, and would die there unless someone found them and turned them around. But this ewe, Daddy said, had been very old. She might have just died anyway.

"How long do sheep live?" I asked.

"Oh, about eight years; that's an old sheep," he said. I could hardly believe it. Why, *I* was eight years old.

In the pasture just below the house, another old ewe, this one with a big lamb at her side, came up to the gate and stuck her nose through the bars, and Daddy sent us up to the house to have Tudy slice up some apples for her. Tilly the ewe had been a bum lamb, raised on a bottle.

That was why she was so tame, Daddy said. We fed her the slices of apple through the bars of the gate, her delicate lips pulling the fruit in. The lamb was too shy to take the food from our hand, so we tossed it to her. The other sheep glared suspiciously at us and moved away, grazing.

We stayed in Wren for a full week, as old friends and relations came to visit and to check out Daddy's new family. They looked us over as though we hailed from another planet. I remember an elderly couple in a Victorian farmhouse, the woman built like a cookie jar, the man rail thin. One day, several round women about Tudy's age, whose favorite color seemed to be electric blue, came to visit. One was an old family friend with the honorific title of Aunt Bessie. She was the mother of a blond paragon named Bruce, now in the Naval Academy, about whose virtues I soon grew sick of hearing.

The grown-ups sat in a circle of chairs in the living room, and the conversation soon languished. I had heard much about Aunt Bessie and her remarkable son, but nothing about her husband. Deciding that somebody ought to get the conversational ball rolling, I walked across the empty middle of the circle of adults and stood in front of her. "Where's your husband?" I asked, coming straight to the point.

Aunt Bessie flung back her head like a startled horse and bleated tearfully, through a tremulous smile, "Oh, honey, he's not here." I retreated, abashed, knowing the mortification of the performer who has just bombed and doesn't know why.

Daddy came into the spare room that night to make sure the lights were out and the covers pulled up. Elizabeth was already asleep, but he stood beside the nightstand, looking at me.

"Those people who came to visit today," he began, "didn't come to see you. They came to see Pappy and Grandma and your mother and me. They don't want to listen to you. After this, you keep quiet and don't speak unless you're spoken to." He turned to go and paused, one hand on the light switch beside the door. "And Aunt Bessie's a widow," he added. "Her husband died about a year ago. You don't ever ask grown-ups questions like that. It's rude." He turned out the light, closed the door, and left us in the dark.

One day Pappy and Daddy went into Corvallis, and took Elizabeth and me along after we begged to go. On the ride in, I leaned up over the front seat and, forgetting Daddy's instructions, spoke before I had been spoken to. "Why do they call you Pappy?" I asked, looking into my new grandsire's ear. To my surprise, both men began to laugh.

"You tell her, son," Pappy said.

"Well, when Uncle Charlie and I were little, we always called Pappy 'Father.' But when we got up in high school, we both turned out to be taller than he was, and started calling him ' Pappy' after the father in the *Li'l Abner* comic strip, because all of a sudden he was the shortest one in the family." Daddy blushed as he told the story, his blue eyes sparkling, and we laughed, too, and I wanted to hug them both, but did not.

In downtown Corvallis, we walked behind Daddy and Pappy as they strolled in and out of hardware and sporting goods stores. When we returned to Pappy's big yellow station wagon, they put us back inside, locked the doors, and told us to wait while they ran a few more errands. The sun, which had briefly lit up the rain-soaked world that morning, had gone again, and the rain pelted down on the wide windshield and turned the neon signs around us to streaks of pink. It was three o'clock in the afternoon, but it seemed later. Elizabeth began to explore beneath the back seat, finding a comic book that evidently had belonged to Michael, and sprawled out on the seat to read it. As Daddy and Pappy turned the corner at the end of the wet street, I felt a wave of panic roll over me. They were leaving us alone here forever; they would never return. The intense fear took over my stomach, my chest. I felt a whimper begin in my chest, then another. Elizabeth looked at me over her comic book and stoutly refused to share my fear. "They'll come back," she told me. Logic, locked deep inside me, agreed with her. Grown men don't abandon perfectly good automobiles in broad daylight. But the screaming panic overwhelmed logic—and me. I *knew* they would not be back. We were abandoned.

We were alone in that car for an hour, perhaps. When Daddy and Pappy returned, they found a bored Elizabeth, leafing through the comic book for a third time, and her tear-stained and frightened older sister, rapidly descending into shame and humiliation at her own inexplicable behavior.

Daddy frowned at me. Pappy ignored us both, absorbed as he was in a diatribe about log exports and how they were driving Woody into bankruptcy. Elizabeth, shifting from concern and reassurance to delight in my shame at one leap, eagerly told Daddy that she kept *telling* Louise that they would be back, but she wouldn't listen and just kept on crying and—

I glared at her and she shut up, but it was too late. Five minutes after we arrived at the farm, everyone in the house knew what I had done. Charlie looked at me sympathetically. Mother, having endured the third degree from Tudy all afternoon about her religion and former husband, was in no mood to sympathize with my morbid fantasies. I wanted Grandmother so bad I couldn't stand it.

I had just managed to calm down a little, when Tudy put a big enamel pot of water on to boil. Charlie and Dorothy had taken Pappy's boat over to the coast that day and brought back a washtub full of Dungeness crabs, which now slowly squirmed and clicked under a blanket of wet seaweed. I crouched down on the kitchen floor and watched them, peering at their working mouths spitting foam and their fierce grasping foreclaws. I held out a sliver of kindling to one of them and felt the satisfying sharp snap as the claw crushed it. The crabs were pale blue and orange on top, dirty white beneath. Charlie carefully picked one up by the back of its carapace, and turned it over. "See," he said, "that's a male, see that triangular piece on the belly?" And he stood up, and dropped the creature directly into the pot of boiling water. One by one, as I watched in horror and disbelief, the rest of them followed, stuffed all alive-o into the cauldron. Some flung out a claw and tried to grasp the side of the pot, their black eyes swiveling frantically on short stalks. I started to cry and to scream something incoherent about not killing them, but Charlie and Daddy only chuckled and kept throwing them in. I turned to Mother, grimly washing dishes at the sink, her back to the carnage.

"It doesn't really hurt them . . ." she began uncertainly, but not wanting to watch the process either as, remorselessly, the last crab was shoved in and the heavy lid clapped on.

Didn't really hurt them? How could being thrown into boiling water not hurt? Couldn't they have been knocked on their heads, or whatever crabs *had, first?* My sobs took on a life of their own, growing to a frantic roar at the injustice of it. Daddy grabbed my arm and propelled me across the living room, past Elizabeth and Michael, who were building a Lincoln Log house on the big round rug, and down the long hall. "Get into your nightgown and get into bed," he ordered.

I lay in the cold dark, disgraced, and heard the others cracking crab at the long dining-room table and laughing. They're laughing at me, I thought, and talking about how stupid I am; and I began to cry, grieving for something lost, and terrified with the knowledge that although he had come back this time, anyone who would throw a live crab into a pot of boiling water was certainly capable of deserting a troublesome scaredy-cat like me.

CHAPTER 12

> But thy throat is shut and dried, and thy heart against thy side Hammers: "Fear, O Little Hunter—this is Fear!"
> RUDYARD KIPLING

Our journey back from Wren ended late on a rainy night in Hilt, as Mother hustled us off to our own beds. I curled up between the icy sheets with my knees pulled up inside my long flannel nightgown and listened to Daddy building a fire in the front-room stove.

I went down to Grandmother's house the next day after school and found her tying flies in the back bedroom. Now that we were gone and Jo was living back in Oklahoma, Grandmother had more time to herself, more time for her own interests and hobbies. This was a side of things I had not considered: that Grandmother might actually *like* having the house entirely to herself in the daytime.

Grandmother, having devoted months of patient labor to Jo's wedding, having cleaned wallpaper and scrubbed walls and polished silver, now had Jo's old room all to herself. Her Singer cabinet sewing machine stood under the window, covered with bits and pieces of doll clothing and tiny patterns of her own design. Beside the sewing machine stood an artist's easel on which sat a half-finished, tentative landscape in oils.

I sat down beside Grandmother and watched her fingers flying around the vise, winding and snipping, while feathers and twill and glue grew into magical insects. Absorbed in her task, she glanced up now and again to smile at me. She was not, I could see, about to stop what she was doing to ask me if I wanted a sandwich. I wandered outside and finally pedaled away on my bike. The world and Grandmother, it appeared, were no longer going to devote themselves to making me happy.

I wanted to tell Grandmother about being left alone in the station wagon in Corvallis, but I couldn't. It just sounded too stupid. I couldn't even explain it to myself. How could I explain the birth of a stranger, who seemed to be myself, growing inside my own head? I was turning into someone else, and I didn't like her very much. The Louise who lived in her Grandmother's house was not afraid of being left alone in a car. But *that* Louise had never been left alone in a strange place. All the way home from Oregon, I was terrified that Mother and Daddy would leave us alone again. I told myself that the fear was stupid, but it was real and pervasive and, as I would discover over the course of the next three years, capable of random mutations that leapt out at me unawares.

That winter I began to wet my bed. Daddy ordered me not to drink water after eight o'clock at night. When I still wet the bed, he ordered me to write, "I will not wet the bed," a hundred times, after school. When I wet it yet again, he assumed I was sneaking drinks of water, lost patience, and spanked me with a thick stick of kindling.

After that, I learned to hide the bed-wetting. I tucked a towel under the mattress, and when I woke in the night and realized what I had done, I laid it over the sheet, slept in my wet nightgown, and said nothing to anyone, especially to Elizabeth, who had developed a habit of tattling on me. Since we made our own beds, I could conceal an accident unless Mother was going to change the sheets that day. Mother sighed and bought plastic mattress covers, and mercifully said nothing, either.

If Daddy heard me go into the bathroom in the night, he would decide that I had disobeyed him and taken a drink after hours and spank me the next day. So I learned to listen for his snores, then slip noiselessly into the bathroom and perch in the dark on the very edge of the toilet, so that my urine fell silently onto the porcelain slope, and not noisily into the bowl. In summer, both Elizabeth and I climbed out the window and peed on the lawn. Best of all, of course, was to sleep on the lawn in the sleeping bags, where we could run a few dew-soaked steps in the night and pee luxuriously under the black shadow of the locust tree.

When spring came, I feared the weekends, feared family outings. If they had to occur, I preferred a picnic on upper Cottonwood Creek, because from there the walk back to Hilt was straightforward and not impossibly long. Walking clear back from, say, Humbug Point on the Klamath River would be hard. On trips to Medford or Yreka, I feared that my parents would desert me as I visited a gas station's restroom, so I went only when I absolutely had to and ran in and out as fast as I could. Once, I hurried out, only to find the car gone from the gas pump. I ran to the curb, looking frantically right and left, then heard Daddy's voice calling to me, irritated. The car, and my annoyed family, were parked at the side of the station, waiting for me. "I had no idea you were such a little coward," Daddy said.

A frog came to live in my throat. I cleared it, again and again, until it became a tic, driving any adult within hearing to distraction. Daddy solved this habit, at least in the car, by reaching back and thumping me a good one on the side of the head, pinpointing my exact location with the amazing peripheral vision that allowed him to see a deer on the other side of the river at sixty miles an hour. I learned to control my throat. Sometimes, when I absolutely *had* to clear it, I would leave the room, leave the house. I thought I was the only person in the world with this problem until years later when Jo, reminiscing with Mother, said, "Remember that awful habit of clearing my throat I used to have?" and Mother glanced over at me as I sat aghast and waited for her to Tell. But she didn't.

I began to sleepwalk in this new house, and when Mother heard me walking through the front room or crashing into the kitchen chairs, she got up and steered me back to bed. In the morning, I didn't remember what I had done. Daddy's mouth twisted when Mother told him about it, and he looked at me contemptuously. But at least he didn't spank me for it.

It was unexpectedly hard to have a Daddy who seemed to have no fears, no weaknesses. He had done things—dangerous things—we could scarcely imagine. He had been a rigging slinger, had topped spar trees on the high-lead side, and then stood up on the stump to look around—two hundred feet in the air. He had set chokers, crawling under giant logs attaching cables, where a log shifting on a slope could crush a man to death. He had seen fallers killed by widow-makers and legs smashed to splinters by a tree that barber-chaired.

He was not afraid of the dark—"There's nothing out there in the dark that isn't out there in the daytime"—and with his remarkable night vision, we were sure he could see this for himself. Yet we, who believed that anything might be out there in the dark, were strangely trustful of Daddy about some things. He was a good driver, for instance, and we were much more comfortable riding with Daddy than with Grandfather, who managed to turn the routine passing of a truck into a kamikaze mission. Daddy was calm in Medford traffic; Grandfather punctuated his Medford driving with profanity and took five minutes to parallel park. Daddy drove at just the right speeds, avoided trouble, and we fell asleep in the back seat of the car at night, serene in the knowledge that he would get us home all right. He could put on snow chains without fuss or curses—which Grandfather could not. Daddy was, in short, competent, and if there is one thing that children recognize, and respect, it is that. I laid the blame for my new sufferings not on Daddy, but on myself.

As I entered the third grade in my first year of our new family life, I had a new teacher. Aristeo Perez was a crew-cut, olive-skinned young man with a bemused expression. His grasp of classroom discipline was far from firm, but he knew the dangers of having a classroom comedian, and he seemed mildly exasperated when I learned to cap his sentences with what I thought were brilliant tags. Still, he smiled at me and seemed to take them in stride. So I was astounded when he gave me a B- in Deportment that first quarter, sullying my unbroken string of As.

Mother learned about it at a parent-teacher conference. "Just wait until your father gets home!" Mother snarled as she led me away from the schoolyard. I spent a miserable two hours waiting for Daddy to come home so he could spank me. This he did with gusto, and it had the desired effect: I never smarted off to Mr. Perez again. He gained some peace and quiet; I lost all trust in him. He didn't seem to notice that last part, or to miss my input at all. Like most adults, he just wanted Quiet.

Down At The Other House, as Elizabeth and I now referred to our lost world, we had rarely been spanked and then only by Grandmother. *Her* spankings were delivered in a white heat of rage for acts of deliberate meanness or disobedience on our part. Daddy was different. He did try other forms of punishment, at first. But my sister and I were perfectly capable of writing "I will not throw rocks at anyone" one hundred times, and then getting in a rock fight two hours later. Physical pain, however, got our attention right away, and Daddy soon discovered that it was a great time saver. We understood that. What we didn't understand, and only learned painfully over the course of several years, was that we could be punished for anything that Daddy thought was an offense, whether we knew it was or not.

When I started having trouble reading the blackboard late in the third grade, Mother took me to see Dr. Lemery, an old and cranky ophthalmologist in Medford. His tone, as he told Mother about my galloping myopia, implied that being nearsighted was a character flaw on my part. Too much reading, he said. Mother yanked my hand hard as she propelled me out of the elevator and back onto the hot pavement.

For the next several years, as my glasses prescription changed every six months, as I broke various parts of the frames in accidents, as I was never without masking tape wrapped around the ear pieces, Mother's belief in the persistence of inherited traits, especially *bad* inherited traits, increased. I was, Mother told me, just like my father: selfish, lazy, a daydreamer. *He* read too much, too. He sat at home and read and read and wouldn't *do* anything. Her lips tightened at the memory, and her brows pulled together until two deep grooves carved themselves between her eyes. "I know someone else who used to act just like that," she told me now whenever I displeased her. She had never said such things to me before, and it stunned me into silence and sent another fear welling up in my chest.

I had not thought much about my father for years. I could scarcely remember his face. We had no photos of him. When I looked at Elizabeth, I saw his slender hands and his pale coloring and light bone structure. Elizabeth and I did not talk about him. We avoided mentioning him to Mother, even going to some trouble not to utter the names "Dick" or "Johnson" in her presence. We were Brannons, now.

My fourth-grade school pictures show a chunky kid with short, lank brown hair, glasses with ugly plastic frames, and a scowl. Although we still stayed overnight with our grandparents every couple of weeks, their house was no longer our home. Grandmother still doctored us when we showed up with a cold, still mixed Capsoleum with Vaseline together and slathered it on our croupy chests, tying a flannel rag around our necks to protect our nightgowns from the grease. She still poured bubble bath into the old claw-legged tub for us, but now she saw, as we stepped out of the tub, the bruises and welts on our thighs and buttocks, and I heard her voice complaining to Grandfather, after she had heard our prayers and tucked us in. I hugged my small, brown stuffed bear to my chest, my talisman against nightmares.

I was relieved when Daddy left early on summer mornings. Mother went back to bed and slept until eight or so, and I could slip into the kitchen and fix my favorite breakfast—peanut butter toast and chocolate milk, eaten out under the locust tree. We enjoyed the summer weekends when Daddy had to work on fire patrol, for then we were free, as we used to be.

Life, we found, was confusing. Elizabeth and I sometimes thought it would be less so if Mother and Daddy would only behave like a normal married couple. But they acted as though they were still going steady. Mother sat right next to Daddy in the car, instead of over by the passenger door like all the other wives in town. After every meal, Daddy hugged and kissed her, and thanked her. They kissed when he left for work and when he came home. On Saturday and Sunday mornings, when we all slept in until seven-thirty or eight o'clock, we heard Mother's giggles and Daddy's chuckles, behind the closed doors of their bedroom.

All those things we had thought cute, once. Now, they were less so. It wasn't that we expected Daddy to be like the fathers on television, those amiable if dim-witted men who perched their daughters on their knees and tossed them up in the air when they came home from work. We had, in fact, no real expectations about that. But Daddy didn't even seem to like us to hug him, and he never hugged us first. Perhaps, we thought, this was because he had no sisters and had grown up without girl playmates. He had, somewhere, picked up the notion that little girls were sweet and feminine and played with dolls, so he was frankly astonished by two hard-bitten hellions who threw rocks, played war and cowboys and Indians, and wanted cap guns for Christmas. Our numerous dolls were used chiefly as props in ongoing sagas of blood and revenge acted out on the back lawn. So Daddy didn't know what to make of us, nor we of him.

That grown-ups never had any fun had long been obvious to us, but now, we felt, Mother was carrying it a bit too far. If she didn't *like* scrubbing floors and doing laundry, then why in the world had she gotten married? I asked her that one afternoon when I came home from school to find her waxing the front-room linoleum. In a flash, the sopping wet end of the waxing rag had slashed across my cheek. "That's all I am to you, an old scrubwoman, is that it?" she screamed at me, and I realized too late that her eyes were narrowed against the light, that she had a migraine headache. Mother's monthly cycles had never been very noticeable before; now they were painfully evident, so that it was dangerous just to come in the back door on some days. When Daddy came home, he spanked me hard with yet another piece of whittled fir.

I kept out of Daddy's way for a few days, knowing that Mother had said *something* to him, and during that time he spent his evenings in his shop—the woodshed—constructing a pegboard and a set of small carved wooden pegs. The peg holes were labeled in ink in Daddy's careful draftsman's script with the names of various household chores. He brought it inside and hung it up on the kitchen wall. Every afternoon when we came home from school, he told Elizabeth and me, we would consult the board and do whatever chores were pegged.

We looked at Mother. The gentle big sister we had known all our lives was gone. This woman, nodding vigorously at the little chore board, was a stranger, and suddenly we had nothing to say to her. But the board had the effect of making conversation unnecessary. We came home from school, we did our chores, then hopped on our bikes to ride the streets.

Often I went down to Grandmother's, to sit under the white poplar, to climb it and lie stretched out on the big branches, as I could not do in the black locust tree of our new home. When I looked up through the foliage, I noticed that some of the very topmost branches had not leafed out, and when I pointed this out to Grandfather, he poked around the base of the tree and pointed out some tiny holes surrounded by frass. Borers, he said. The tree was full of them and probably ought to come down. Several years before, he had planted a couple of maple saplings on the other side of the yard, as though in anticipation of the old tree's demise.

"Nonsense," said Grandmother, when I told her. "It just needs water." And she moved the hose to the dry corner by the fence and turned it on full force.

In the backyard, I climbed the cherry tree, carrying an apple and a comic book, and sat hidden among the branches for an hour at a time. I sat cross-legged on Grandmother's front-room floor, in front of the glass-doored bookcase, before the solid unchanging comforting old volumes. I read Zane Grey and Jo's old Nancy Drew mysteries and *The Robe.* I read Howard Pyle's tales of Robin Hood and *The Swiss Family Robinson.* I discovered *Penrod* and *Tom Sawyer* and *Huckleberry Finn.* I devoured Terhune's stories of noble collies and absorbed the Greek mythology in Hawthorne's *Tanglewood Tales.* I read as much as I could at Grandmother's, because I wasn't allowed to read at home anymore. Dr. Lemery had told Mother not to let me read outside of school. This threatened to deprive me of my chief drug and comforter. And after Daddy caught me coming back from the library one afternoon with a book under one arm, I didn't dare go *there* any more.

Our ballet lessons ended shortly after Mother's marriage, when Karen Adele married a Forest Service engineer, and so neither Elizabeth nor I ever "went up on our toes." With Jo also married and gone to Oklahoma, our musical careers ended abruptly.

Sometimes I forgot what I looked like now and thought of myself as that pretty, cheeky, graceful seven-year-old with the long hair and strong soprano voice and the ability to learn any song or dance in ten minutes. But when I tried to act like that girl, Daddy was there, to tell me that I was fat and wore glasses, which I had just broken and which would now cost him money to have fixed. I looked at snapshots taken only two years before in disbelief. That child with the confident grin, tying her shoes on the couch, while Grandmother stood behind her at the old buffet, writing checks, the morning sunlight spilling over her shoulder—that couldn't be me, couldn't be the bespectacled fourth-grader with the ugly haircut and glasses awry on her nose.

"There was always something different about Johnny, right from the start," his brother told me many years later, looking at me with the blue Brannon eyes that were so like Daddy's. As I began to learn those differences, I also began to understand why he had so little patience with us, we children of privilege who demanded endless affection, although we had never had to work a day in our lives to earn it.

Daddy was three years old when he was diagnosed with leukemia. After weeks in a San Diego hospital, a series of massive blood transfusions, and several life-threatening infections, he was alive, but his parents were broke. The family moved out to a farm near Lakeside, where Al raised beans and chiles and ran a well-drilling business.

At harvest time, Al hired Mexican field hands who brought their families, everyone working together. Al let them pick chiles for their own use and watched as the young men picked ditch weed—wild marijuana—and rolled the dried leaves in cigarette papers and smoked it. They were too poor to buy tobacco. One day Pappy looked at the six children walking behind one of the hands and asked him in Spanish, "Why do you have so many? How do you expect to raise them all?"

"Oh, but we don't expect to raise them all, sir," the man answered. "Some will die. *That's* why we have six."

Behind Pappy, Johnny stood barefoot in the furrow and listened.

The boys went barefoot most of the time in the California warmth and lived outdoors most of the year, sleeping on a screened-in porch. And to the astonishment of the doctors, Johnny slowly got better, although Charlie, a year younger, quickly outstripped him in height and weight. Until his teens, Johnny was never really robust, and he caught every available sickness—measles, mumps, chicken pox, whooping cough, shingles. He grew slowly and stayed thin and often caught colds, while Charlie was a husky kid who was almost never sick.

Lakeside was rural then, but the streetcar lines ran all the way out from San Diego. For a nickel, the boys could ride the cars into the city to visit Tudy's brother Hank, but there were few nickels. The boys both worked whenever they weren't in school. They hoed beans and drove the cultivator horse and milked cows. They helped drill wells and watched as Al or their Grandfather "Pop" Brannon "witched" a new well location with a forked willow branch.

In the countryside around the farm, bobcats and coyotes and occasional cougars roamed, and snakes abounded. The boys ran trap lines and sold the furs of foxes and raccoons. They killed crows and magpies with slingshots and later with .22 rifles and turned the heads in for the ten-cent bounty.

When Daddy told us stories of those times, I learned for the first time that this fearless man had once been afraid of something. Until he was about twelve, Johnny was terrified of snakes, yet he never saw a snake on any particular day unless he had dreamed of one the night before. If he had dreamed of one, he tried to avoid snakey places, but it did no good—he *would* see one that day. When he was twelve years old, the fear of snakes abruptly left him and with it the dreams of them. After that, he often caught snakes and took them to the San Diego Zoo to sell, carrying them to the city inside his shirt, wrapped around his torso or curled up in a pocket.

Pop Brannon owned some land in Oregon; he and Grandma Rosa moved up there just before Pearl Harbor. When Pop's health failed in 1943, Al and Tudy sold the bean farm and moved to Oregon, too. By the time Pappy and Tudy bought the property near Wren and invested in a few dairy cows, their cash was gone, and they were reduced to buying groceries with the cream check. Pop and Rosa lived in Corvallis now, where Pop complained endlessly about the indoor plumbing. It wasn't clean or healthy, he declared. "Not right, people shitting in their own houses," he muttered.

Pop's perspective on Need versus Luxury had been shaped in a hard school. When he was thirteen years old, his entire family had moved away and left him alone to prove up on a homestead in Nebraska. He lived by himself on the prairie for four years.

Pop's father, Bushrod, brought his wife Elizabeth and their eleven children to western Nebraska in the 1880s. He took up a standard quarter-section homestead, but he had no real intention of staying. All around him, cattlemen were buying up homesteads to control the water sources and squeeze the homesteaders out. In fact, the standard way of doing business in that country was to prove up on a homestead after seven years, then immediately sell out to one of the big cattle outfits and move on. Three years after filing on the homestead, Bushrod had an opportunity to buy some land in Oregon. Like the Roushes in eastern Nebraska, Bushrod was already sick of the drought and grasshoppers and blizzards, but he was not about to walk away from 160 acres on a creek. So Pop—young John—was left behind to maintain the Brannon ownership of the claim and to farm it. For four years, alone, he plowed and planted and harvested, renting teams from the neighbors for the heavy work, raising what he could, selling what he could, and eating the rest. After four years, as planned, he sold the claim, put the money in a money belt, and began to walk west. He did not hurry. He hitched rides on wagons when he could. He avoided the railroads, both from a fear of being robbed and to resist the temptation of buying a ticket with some of the precious money. Months later, he walked into his parents' house in western Oregon. He was just seventeen, and his life so far had taught him that land was expendable, that you could and should leave it and walk away and find more land and better when the opportunity offered. His years alone had taught him that loneliness was normal, that a man should be able to live without family, without company, without entertainment, and without complaint.

Pop's Great Walk West was the central myth of Daddy's childhood, the story told over and over, the mark at which the Brannon men aimed their conduct. Its lessons were the lessons of self-sufficiency and thrift, its hero a man who learned to live alone and to conquer whatever emotions—fear, loneliness, sorrow—dwelt with him in that sod house on the plains. When he walked away from the claim, the legend told us, he walked away from those feelings and into a manhood devoid of civilization's foolish false needs. And so, the moral of it went, should we all.

CHAPTER 13

> Open my heart and you will see
> Graved inside of it, "Italy."
> ROBERT BROWNING

Italo Marin pulled greenchain *for thirty years at the lumber mill in* Hilt. He went to work in the mill at sixteen, where his father, John, had gone before him, where his son, Frankie, would follow.

That a man could spend thirty years pulling sawn boards from a conveyor belt, stacking them in piles according to size and grade, and be not only content, but positively happy, may be hard to credit, but it was true, and when the mill closed forever in 1974, Italo said he knew it would happen eventually. He just hadn't thought it would be this soon.

Something about Hilt compressed time and made thirty years run past in a moment. Perhaps Old Man Marin, Italo's father, felt it, too; felt himself still that young immigrant from Italy, glad of a job in a lumber mill, even as his grandchildren's contemporaries saw him as a strange old man who lived in a tiny house and, almost blind, sat on a wooden bench beside his front door, shouting songs in Italian into the sunshine. His wife puttered endlessly in her kitchen garden in a long black dress and an untidy bun, squinting at us and murmuring incomprehensibly when her granddaughters came to visit with their friends. We all drank Kool-Aid in the microscopic kitchen beneath the bundles of sage and oregano hanging from the ceiling while Nonno—as children called the old man—sang Verdi out front.

Across town, Inez Marin, wife of Italo's brother Tony, lived rather differently. She wore slacks and a permanent wave and talked about the election, while her daughter Lynn sat at the kitchen table with my sister Elizabeth, leafing through a fashion magazine. Like many third-generation Italian kids in Hilt, Lynn spoke only a few words of Italian herself, since Tony and Inez only spoke the language when they didn't want the kids to understand. "I'm an American. The Pope doesn't tell *me* what to do," Inez declared through a haze of Chesterfield smoke, leaning back against her kitchen drainboard. "Tell your grandfather that," she said, grinning without malice at Elizabeth, who had just told her that Grandfather said all the Wops would vote for Kennedy because he was Catholic. Then, looking over Lynn's shoulder at a photo in the magazine, she changed the subject.

"Jesus Christ! Skirts are getting so short, pretty soon the gals'll be wearing two hairdos!"

Inez did not believe in censoring her opinions to suit the ears of children.

When Elizabeth repeated the substance of the conversation to Mother, she laughed at the fashion prediction and was mortified that Elizabeth had repeated Grandfather's embarrassing opinions. Meeting Inez at the post office the next day, Mother apologized.

"Oh, Barbara," laughed the former Inez Catuzzo. "Sure, I'm a Wop. I'm proud of it! I know what Bill Roush thinks, and it's a free country, so it don't bother me a bit. I know you don't think that way and don't teach your kids that, either, so don't worry about it."

Over the years, however, Grandfather got along better with the Italians than with most people in the box factory. When Grandfather invited Primo Favero over to see his new washing machine in the late 1940s, Primo listened respectfully to the presentation, went home, and thought about how to repay this evidence of favor from the boss. He came back another day, bringing a bottle of his homemade wine, and sat talking about food and gardens, until he found himself assuring Grandmother that his wife, Mary, would show her how to make ravioli. Primo went home and told this to Mary, who exploded. Her ravioli were almost sacred, a family secret, handed down from mother to daughter, jealously guarded. But in the end he talked her around, and Mary walked down to visit Grandmother and found only a shy woman who also knew what it was to be alone in a strange land. Mary wrote down a list of ingredients for Martha to find and promised to bring the asiago cheese she would need.

Making ravioli turned out to be far more than a recipe, which in any case existed only in Mary's head, just as she had learned it from her own mother. It was a two-day operation, during which Grandmother had to translate Mary's handfuls of this and pinches of that into cups and teaspoons. But by the time the dough and the filling and the sauce were at last completed and the first batch of little filled pasta squares boiled, and the sauce spooned over them, and the whole creation topped with grated asiago cheese and placed on the table in the big platter, the two women were friends.

Grandfather's Italian employees never argued with him, never glared at him, and showed little interest in unions, so the Company liked them, too. The Favero boys used to bring their father a bottle of cold home-brewed beer every day to the box factory, hiding the bottles in the bundles of box shook in the factory warehouse. After a while, there was always an extra bottle for the foreman if he wanted it during the three o'clock break. Bill knew the bottle was there, but he left it alone until Friday afternoon, then carried it home in his inside jacket pocket. What the Company winked at in a factory hand, it would not tolerate in a foreman. Foremen had been fired for even a suspicion of drinking on the job.

Until the 1950s, most of Hilt's Italian families lived in Little Italy, on a hill west of the railroad tracks. They ordered grapes by the carload and made wine together, all through Prohibition. The Company made sure the county sheriff was looking the other way as wagons hauled the grapes up the hill and children stomped them in big vats. The grape pulp was distilled a second and even a third time to make grappa in stills hidden in the brush. Mr. Hunt, the town constable, warned the Italians if a federal raid was imminent.

Most of the Italians arrived in Hilt just after World War I, although a few were there before. They built their own houses, with cellars for storing vegetables and canned goods and cheeses. They raised chickens and vegetables and cooked on wood-burning ranges, long after everyone else in Hilt had electric stoves. By the 1950s, the second generation bought their wine at the store and lived scattered throughout the town, like everyone else. Since there were no longer any children there in our time, Elizabeth and I seldom went to Little Italy, but sometimes when my classmate Linda Zanotto and her sister Laura went there to spend the day with their grandmother, we tagged along. We stayed for lunch, which was always spaghetti and sauce, homegrown vegetables, bread, and red wine. The bathroom was simply that—a small room with a bathtub and sink. The toilet was an immaculate outhouse behind the garden, with a pine seat scrubbed white with lye and a stiff brush.

Old Mrs. Zanotto's garden was surrounded by a high fence of posts and chicken wire, to keep the deer and jackrabbits away from the zucchini and peppers and beans. The front of the house faced east, looking down on the box factory, and it had something that we thought a delicious novelty—a cellar, dug under the front porch. We could walk into it through a short door beneath the front porch, and enter a room of marvels. Giant dried cod hung head down from the beams, next to hunks of dry cheese in netting and sausages shrouded in white skins.

Out in the garden, we picked young rhubarb stems and dipped them in handfuls of sugar filched from the bowl on the kitchen table. We gathered eggs from the chicken house and were pecked by outraged hens.

Hilt's first Italian families were poor; just how poor was by our time almost forgotten—by non-Italians. But Mother remembered arriving at an Italian schoolmate's house once at suppertime and finding the entire family dining on spaghetti. Just spaghetti and well-watered wine. No sauce, no cheese, no bread. The family could have used the Company's credit chits to get food from the store until payday, but they had a horror of debt. It was better to go hungry.

Cappellos and Michelons, Trincas and Marins, Cunials and Alphonses. Lena Foggiatto, who was our baby-sitter when Mother and Grandmother and Grandfather went to dances or lodge meetings; Mario Michelon, who grew up in Hilt, went to college, and returned to Hilt as the school principal. Mario was cheerful and black haired, with Buddy Holly glasses, and full of optimism about community projects. He organized a contest among the school children to name the streets of Hilt—a decades-long oversight that he felt needed correcting. Some of the streets had long had nicknames—Front Street, Church Street, Adobe Street—but others did not.

There were prizes and considerable excitement in school for a while, and the Company promised to make signs for the renamed streets. But Front Street wasn't Siskiyou Avenue, however good that sounded, and it was no use pretending it was. The signs fell down after a while, and no one put them back up.

One Halloween, the high school boys, following sacred tradition, sneaked into the alley behind Mario's house, took all the wheels off his car, put it up on blocks from the woodpile, then thoroughly waxed every window on both car and house. The next day at school, our principal seemed remarkably ill humored about this. He called a special school board meeting, managing to ram a resolution through forbidding trick-or-treating the following year. The PTA organized a Halloween carnival, and we wore costumes and bobbed for apples in the lunchroom, and hated Mr. Michelon.

My sister, discussing the matter later with Inez Marin and Lynn, reported that Inez said that spoiling trick-or-treating for the kids was pretty damned intolerant of Michelon, considering what he and his buddies had done to all the outhouses on Little Italy one year. "Served him right that those hoodlums put his car up on blocks," she chortled. "Since he went off to college, he's just plain lost his sense of humor."

CHAPTER 14

In plains that room for shadows make
Of skirting hills to lie,
Bound in by streams which give and take
Their colors from the sky.
RALPH WALDO EMERSON

Hilt Creek was a seasonal stream, born in a draw far up the steep slope we called Skunk's Peak. Swollen by winter rains, it hurried across a dry, eroded flat, slid under a fence and through a galvanized culvert, and emerged into a meadow below the new school, where it meandered through sedges and grasses, past a few scrubby willows and swathes of red curly dock, and relaxed for a little while.

Here, the creek could meander thanks to a buildup of sediment washed down from the soft shale hills. Twenty inches of rain a year were not enough to flush the sediment off the flat, and so deep soil built up here. Above the fence on the steeper ground, twenty inches of rain a year were just enough to ensure continued gullying, for throughout the spring, until the annual grasses dried up, the cattle of the SS Bar Ranch grazed there, trampling and denuding the streamsides.

The foothills and valleys around Hilt are an ecological extension of the great Central Valley—a savanna, in effect. Since the 1860s, the dominant plant communities have been based on introduced grasses and forbs from the Mediterranean basin—wild oats, filaree, cheatgrass, annual bromes and fescues. During that decade, a years-long drought coupled with the unrelenting assault of millions of cattle, horses, and sheep completed what the early colonists from Spain and Mexico had begun.

By midsummer, Hilt Creek was a dying series of stagnant pools, but in the first warm days of April, it was still a stream, and long strands of black toad eggs, strung together by lengths of transparent jelly, appeared along its banks. Clinging to rushes beneath the overhanging banks, blobs of Pacific tree frog eggs floated. Western toads clasped each other in catatonic affection in the shallows; gravid female tree frogs hopped through the wet grass, pursued by much smaller males. Meadow larks rocketed up from the dead grass or trilled from fence posts; red-winged blackbirds celebrated their territory on every willow. Water skippers fled across the quiet waters, and riffles gurgled, the color of rum pudding sauce, in the wet and fecund California spring.

In the plunge pool below the culvert, foothill yellow-legged frogs hid on the muddy bottom, perfectly camouflaged. The meadow became a cacophony of peeping and croaking, which grew silent as I passed up the creek, then resumed behind me. I caught the singers, picked them up, let them go again. I scooped up their eggs and tadpoles in coffee cans and took them home and attempted to raise them, with varying degrees of success. The best way was to empty them into an old dishpan sunk in the mud below the faucet on the east side of the house. Algae colonized the sides; the toad eggs became little black commas, then tadpoles with quickly flicking tails, growing larger, and finally sprouting legs in August. Their mouths widened and their eyes popped up, until finally one morning they sat on the big rock in the middle of the pan, absorbing their own tails. The tree frog tadpoles turned green, the toads a mottled brown, before they climbed out and hopped away into the iris beds.

Below the meadow, Hilt Creek flowed across the alley and slid into a long series of cedar culverts that carried it through the middle of town and under the railroad tracks, disgorging it on the west side of town, into a deep sluggish swamp of cattails that eventually drained into Cottonwood Creek. Here, in an inaccessible morass of mud and algae, bullfrogs lived, squeaking and leaping away into the safety of the deep slime at the first sign of movement along shore. Bullfrogs are not native to the West Coast states, but they adapted happily to the ponds left by gold dredging along the Klamath River and migrated up the creeks in the wet season.

The gold miners who followed Cottonwood Creek to its sources in the 1850s found little gold. The richer deposits lay beyond the granite peaks that separated Cottonwood from the vast Beaver Creek drainage. The soil of Hilt was not disturbed, for it had no gold or mercury or copper like that of the serpentine formations of Red Mountain or the folded blue schists of Condrey Mountain. Here there were only shale and sandstone. In Hilt, we looked up at the alpine slopes below Mount Ashland, where rich meadows and groves of timber waited for the spring and its onslaught of range cattle and chainsaws. In our arid little valley, spring brought cattle to roam the streets and vacant lots for a few days, before they were driven past the barbed-wire fences at the edge of town. So the little meadow was mostly undisturbed by hooves and teeth, and the frogs flourished.

In Grandmother's yard, I had always followed insects and spiders through the grass and peered into birds' nests and into the golden eyes of the big toad who lived a sedentary life under the water tap on the side of the house, but now all nature began to exert a powerful fascination for me. I had always known there were ants, but now when I stood out in right field in the schoolyard, they were interesting, and I squatted down to look at them scurrying in and out of their holes. One day another fielder yelled "Hey, Nature Girl!" as a ball was finally hit in my direction, and the name stuck.

When Dr. Lemery examined my eyes just after the start of my fifth grade year, he was so alarmed at my advancing myopia that he told Mother that I should only go to school for half a day and should stay outside as much as possible. The heavy subjects were all in the morning, so I was excused in the afternoons. Not long after that, several boys began throwing rocks at me if they met me after school. At first I simply threw rocks back, but after I got caught at it and was spanked, I avoided them instead.

Being out of school when everybody else was in it was no privilege, despite what the boys thought. I couldn't go to the store or the post office or to Warrens's; it was too humiliating to explain to curious adults—and they *would* ask—why I wasn't in school. The chief deterrent to misbehavior in Hilt was simply the impossibility of hiding much from the neighbors; this applied equally to a friendless bum walking the railroad tracks, or a child riding a bicycle around town at two o'clock in the afternoon on a school day. I started riding my bike to the edge of town, crawling under the barbed-wire fence, and hiking behind the little ridges, unseen. I came home with pockets full of interesting rocks and asked Daddy what they were. He told me and started to bring ore samples home from old mines in the mountains. I put them in cigar boxes and labeled them. I found books at school about rocks and fossils and borrowed Mother's garden trowel to dig in dry gullies, looking for arrowheads. I started a flowerbed and my own vegetable patch in a corner of the garden.

I climbed up Watertank Hill and walked around the two huge, red water tanks, and when I wondered aloud how much water they held, Daddy wrote down the formula for the volume of a cylinder and told me to find out for myself. I hiked back up with a yardstick and a piece of chalk and worked my way around the wooden walls and figured it out.

I started collections: of rocks, of pressed flowers, of lichens and seeds and the pieces of purple glass scattered on the hillside behind the old hotel. Books about the plants and animals of the West Coast were unknown at school; the few books available described eastern species like leopard frogs, but were silent about the species I knew. I brought home paper and crayons and drew and wrote about frogs and birds and insects and folded the pages and made books of them, with crayon illustrations. When I didn't know something about an animal, I either extrapolated based on my own observations, or made facts up out of whole cloth. The results were hilarious, but not so different, I later discovered, from those of Pliny in a similar situation.

I caught lizards and snakes among the rocks and sagebrush: bluebellied western fence lizards, which never bit, and big greenish tan alligator lizards, which did. Some of these were almost a foot long—and hissed. Rather to my surprise, Dad built a cage out of wood and wire mesh for one of my captured lizards. All summer I fed it live insects; in winter it burrowed beneath the gravel on the bottom of the cage, and we placed him under the house to hibernate.

The more I learned about the lives of animals, the more I liked them and identified with them. One morning, I stood with a group of children in the school parking lot, watching as a ranch hand stuck a rifle barrel out the window of a pickup truck, parked over on the county road next to Bob Trinca's gas station. The rifle belched fire at a coyote in the field just beyond the schoolyard fence. The coyote went from a trot to a flat-out streak, while bullets kicked up dust around him and the sharp cracks of the gunshots echoed off the brick wall behind us. By the time the coyote disappeared into a gully, safe for the moment, we were cheering for him.

My parents unwittingly encouraged this empathy for animals by subscribing to a series of nature periodicals published by the National Audubon Society. Once a month, a colorful magazine arrived, accompanied by a flat of gummed photo-stamps, to be pasted inside. The Audubon people took a sympathetic view of coyotes, pointing out that they consumed harmful rodents and were part of the balance of nature. But one issue in particular was an eye-opener for me. In *Our Vanishing Wildlife,* I learned for the first time that we human beings had entirely destroyed some birds and animals. I looked at pictures of the heath hen, Carolina parakeet, passenger pigeon, Labrador duck, Steller's sea lion, great auk, and dodo. Sixty million bison, slaughtered for their hides and tongues. Other creatures, barely holding on—bald eagle, whooping crane (then down to twenty-four individuals), kit fox, Key deer. I was appalled and saddened.

My concern for animals was selective, of course. The frogs I caught and the lizards that slid away into the rocks leaving me holding their wriggling tails undoubtedly regarded me as just another predator, if a less lethal one than the boys who bombed frog pools with boulders and tormented garter snakes until I paid them fifty cents to stop and give them to me. I loved to inspect the flattened bodies of jackrabbits killed on the paved county road and turn them over with sticks to look at their dusty eyes and squashed green intestines. I pried the desiccated corpses of toads up from the hot asphalt and sailed them into Sharon Martin's yard just to hear her scream. But above all, I was fascinated by the deer carcasses that arrived in the back of Daddy's pickup during the deer-hunting season.

The Pacific blacktail deer was the only wild ungulate to survive the perils of the nineteenth century in western Siskiyou County, almost the only big game animal to survive at all. Only the black bear, shy and wary, and the very rare mountain lion, on which there was still a bounty of fifty dollars, coexisted with it. Wiped out entirely were the Roosevelt elk, California bighorn sheep, California grizzly bear, and gray wolf. In eastern Siskiyou County, in the lava rimrocks and juniper forests, lived the mule deer, of which the blacktail is generally considered to be a subspecies. Mule deer bucks can weigh over 300 pounds, but a 150-pound blacktail is very large, and most of the bucks we saw were smaller.

Blacktails were not only smaller than mule deer; their ears were not as big, and the top of their tail was all black, instead of white in the middle. In areas where the ranges of mule and blacktail deer overlapped, as they did around Hilt, hybrids occurred, marked by their heavier antlers and a narrow black stripe running the length of the tail.

To me, the bucks were huge, and beautiful, whether they still wore their reddish tan summer coats, or had already shed it for the plushy bluish gray winter pelage. Their eyes glazed and protruding, the bucks—two of them each year between mid-September and the first week of November—arrived in the back of Dad's pickup truck, gutted, their bloody hearts and livers wrapped in old pillowcases beside their antlered heads. I watched the ticks crawling out onto their dry black noses and flipped them onto the lowered tailgate and crushed them with a splinter.

Although Daddy sometimes went hunting with one or two other men from work, he had no close friends in Hilt. He seemed immune to the male bonding rituals of other local men. He belonged to no lodge, did not drink beer at Warrens's, and when he bagged a buck, he didn't show it off at the Company store. He came home with it as soon as he could, without stopping. Most of the time, he hunted alone.

His other hobbies were solitary, too. He melted lead in a long-handled iron ladle, over a bed of coals in the front-room stove, and poured the liquid metal into molds for fishing weights. He reloaded his own hunting ammunition, setting up the scales and presses on the kitchen table after supper. He chipped, ever more skillfully, at projectile points and knives of obsidian or glass. He took Elizabeth and me out to the old ballpark in the evenings, and taught us how to shoot, using a .22 Remington single-shot rifle with a cut-down stock. Sometimes he took us on short evening hunts during deer season, and we learned to sit quietly on a hillside for an hour or so, listening, while evening fell through the forest. Daddy never shot a buck on any of these short trips, but we felt that even to Daddy, what was really important was the sitting, the listening, the sight of a doe slipping cautiously by on a trail below us, or a jay swooping down, giving away our position with its shrieks.

Daddy also bought a hunting rifle for Mother, a .222 with a scope, the smallest legal caliber for deer. He mounted a soft rubber shoulder protector on the stock, to minimize the already small kick, and taught her to shoot. Nervously but meticulously, she stuffed her ears with cotton and proceeded to learn. She became rather good at it, although the last thing she really wanted to do was to shoot a deer. The real purpose of having another hunting license in the family was to have more deer tags available. It was quite common in Hilt for men to fill their own tags and those of their wives, too. The wives always had a license and deer tags and knew how to shoot a deer rifle. If the male hunter shot more bucks than he had tags for, he would hide the extra carcass, drive quickly back into town, pick up the missus, and drive back out with her. She would fire a shot or two into a stump and place her tag on the buck's antlers. This was, of course, illegal as hell.

Before hunting season began, Dad took the deer rifles out in the woods to sight them in. He found a large open area and set up a box about a hundred yards away, with a target tacked on it. Resting the barrel of the rifle on a folded coat or blanket on top of the hood, he opened the action and slid in a cartridge. While we waited for what seemed an eternity, wads of cotton stuffed in our ears, he fired a group of three shots. When he put the rifle down, we ran ahead of him to check the target. He circled the bullet holes with crayon, then walked back to the truck and repeated the process, adjusting the scope or the open sites each time, until he had a one-inch group in the center. When he was satisfied with the performance of the deer rifles, he pulled out our .22 rifle and let us shoot.

We learned to squeeze the trigger, not jerk it. We learned to shoot offhand—from a standing position—and sitting down, our elbows braced against our knees, and finally belly-down in the dirt. We learned to use the truck hood or a tree to brace against for a steadier aim. We learned to relax and let the sights drift across the target, squeezing the trigger a bit more each time, until the rifle fired. "You shouldn't know exactly when the gun will go off," Daddy told us.

During the deer hunting season, the Company store manager opened up on Sundays if a hunter wanted to hang up a buck carcass in the walk-in cooler, and many did this, paying the butcher to cut and wrap the kill. So when kids saw a pickup truck or a jeep parked at the loading dock on the store's west side, we ran up to stand around and stare, while the hunter and his friends drank beer and sat on the edge of the dock, or leaned against the truck bed, seemingly in no hurry to get the buck out of the hot sun to be skinned. That, Daddy said, was one reason a lot of people said they didn't like venison: if you treated a beef that way, it wouldn't taste good, either.

Daddy was a fine shot and a capable amateur gunsmith. He turned half of the woodshed into a shop. His grandfather's tools hung above the workbench. Guns and the tools needed to build and shape and repair them were really the only consumer goods in which he was interested. Cars and trucks were just transportation. In the evenings, he read *The Shooter's Bible* and *The American Rifleman.* He never owned a gun cabinet; he didn't believe in showing off his guns, in having them out where casual visitors could see them. For the same reason, he never put a gun rack in the rear window of his pickup. The rifles were hidden deep within the bedroom closet; the .44 revolver lived in the top drawer of his bureau. Elizabeth and I knew where they were, and we also knew that if we ever touched them without his permission, he *would* kill us.

As the woodshed-workshop steadily filled with Daddy's possessions, many of them brought from Wren, or mailed in large wooden powder boxes, I noticed a row of spike antlers hanging over one of the roof beams. I knew, since I had assiduously read the California big-game regulations handbook, that it wasn't legal to shoot spike bucks in California. When I asked him about them, he looked up from the gunstock he was sanding. They were Oregon bucks, he told me, where it *was* legal to shoot spikes and sometimes even does. And that was true.

I wasn't the only one to notice those little spears of bone. They started something in the suspicious mind of Bill Straight, the game warden. Bill wore black plastic glasses and tried to comb his thinning hair so that it covered his bald spot. He used to hang around the Company store in his State Fish and Game Department uniform on Saturdays, smoking and gossiping with returning hunters. One day he saw Daddy's truck come over the railroad tracks with a buck in the back and followed him up to the house. As he lounged over to the open back door of the woodshed, he could see that the deer tag had already been validated by Hank Mostovoy, a Forest Service employee who sometimes hunted with Daddy. Although Daddy didn't seem anxious to make conversation, Bill stayed to talk and admire the buck and to ask where Daddy had shot it. "In the neck," Daddy said, deliberately misunderstanding him.

Straight threw down his cigarette in the alley, ground it out, and turned to leave. As he got in his truck, he cast a final glance upward through the open door at the row of spike antlers.

"Where *did* you shoot it?" I asked, as Daddy sat on the woodshed steps, sharpening the blade of his pocketknife, preparatory to skinning the buck.

"West Fork of Beaver Creek," he told me, his eyes lighting up with sudden amusement. "But Bill Straight doesn't need to know that." He spit on the gray whetstone.

I nodded and sat down beside him, watching the knife blade describing circles on the small square stone. I was rapidly learning that there were no guarantees in life, but just for this moment, Daddy and I were allies. There was something about Bill Straight that neither of us liked.

Once a buck was hung up by his antlers in the woodshed, we watched as Daddy delicately removed the skin with his knife, slicing carefully through bubbly connective tissue, never nicking the hide. Between hide and meat, the bloodshot tissue from the wounds spread, and he traced the path of the bullet from the small, blackened entry wound to the massive exit wound, where chunks of rib or shoulder blade often shattered into shrapnel. He pried the mushroomed bullet out of the flesh and examined it, showing it to us, to let us know what a bullet could do. Sometimes, if the buck had been gut-shot before the fatal heart or lung or neck shot, Dad spent a long time with a damp cloth, wiping the splatter of liver or rumen off the walls of the rib cage. Finally he pulled a cotton deerbag up over the carcass and tied the strings securely under the buck's jaw, to keep flies away.

The deer hung for three days—longer if the weather was cold. Before he left for work, Daddy draped the carcass with several heavy wool blankets, to keep it cool. At night, he took them off, allowing the carcass to breathe. Sometimes I went out into the woodshed in the dark and slapped the stiffened body gently, wondering at the hard, hollow sound of it. The head stared up at the rafters, dried tongue protruding, the hairy ears hard as rock.

Daddy cut up his deer himself, after supper on the kitchen table, with Mother carefully double-wrapping the cuts in butcher paper and taping and labeling them. He trimmed as much of the fat off as possible, since deer fat goes rancid and leaves a gamey flavor in the meat. We put the meat in the freezer and ate the heart and liver first, then a package of tenderloin, sliced crossways and rolled in flour and salt and gently sauteed. Daddy sawed off the top of the skull and hung it and the attached antlers over a rafter, with the tag wrapped around a tine.

I had no problem with killing and eating deer, or trout, or rabbit, or grouse. What people liked to kill and eat, I reasoned, they would take care to conserve, so that there might be more next year. The deer hunters of Hilt were no threat to the tribe of deer as a whole; no threat to the deer that every night walked our outer streets. It was the other animals, the ones nobody liked, that I worried about. Cougars and coyotes, hawks and badgers, frogs and garter snakes. Shot, trapped, killed, tormented simply because they were alive and available, and it was fun. Bounties on cougars, ranchers shooting every coyote they saw, these things made me angry, and in this anger I seemed to be alone.

CHAPTER 15

> Its vanished trees . . . had once pandered in whispers to the last and greatest of all human dreams; for a transitory enchanted moment man must have held his breath in the presence of this continent, compelled into an aesthetic contemplation he neither understood nor desired, face to face for the last time in history with something commensurate to his capacity for wonder.
> F. SCOTT FITZGERALD

Daddy always packed his .44 revolver in the car or truck, in the glove compartment or under a coat on the seat. One Sunday morning as we drove along a logging road west of Hilt on our way to a patch of blackcap raspberries that Daddy had found, the car stopped suddenly, and I found myself looking into the eyes of a coyote pup standing by the side of the road. Perhaps three months old, it still had the big paws and silly expression of babyhood, and its mouth was stained red from the thimbleberries it had been pulling down and eating, one at a time. Slowly, Daddy reached over and opened the glove box, pulled the revolver out, and slid it free of the holster. Slowly, he extended it forward, until it rested on the frame of the lowered window. The pup was staring at me and seemed not to notice the movement.

I had been riding with my window down and my chin resting on my crossed arms. Now, as I heard the hammer clicking back and realized what Daddy was about to do, I slapped my palms hard against the car body. "Shoo!" I shouted. The pup melted into the thimbleberry stalks and was gone. Daddy let the hammer back down. Without a word, he put the revolver away and the car in gear and drove on.

Later that day, sated with blackcaps, the trunk loaded with buckets of them, we picnicked on the West Fork of Beaver Creek. Chipmunks peeked at us around enormous tree trunks, and as I held out a crust of bread, one of them made a dash for me and stood up on his hind legs to snatch it out of my fingers. Daddy made a mock lunge at the little rodent, and it squeaked in fright and retreated. "Shoo," he said, looking at me.

I felt my face getting hot, but I stared back at him. "He was just a pup," I muttered, my heart hammering. "He wasn't hurting anything."

I knew he was displeased, but I also knew, somehow, that even Daddy wouldn't hit me for scaring the pup away from his gun. Now, he only shook his head. "You have got to get over this complete sympathy for animals," he said, and turned his attention to the fried chicken.

My appetite was gone. I was learning, at the age of nine, just how inconvenient a thing a conscience could be.

Grandfather did not hunt deer, or anything else. He did own an old double-barreled twelve-gauge shotgun, which he had bought twenty years ago to go pheasant hunting with a fellow Mason, but he had never hunted since. Once a year he brought the shotgun out of his closet, cleaned it, and put it back. Grandfather did not own a pickup, and since he would not have risked his tires and the Oldsmobile's paint job on the rough logging roads, Daddy was our introduction to the mountains and the forest.

Daddy liked to cut his own firewood. Parking the pickup a safe distance away, he carried his large and unreliable McCulloch chainsaw up to a snag or double-topped green tree, felled it, bucked it up into rounds, and loaded them onto the truck.

We loved to watch him work, hard hat perched on his head, neck craned back to watch the top of the tree as he pulled the chainsaw's bar away from the final cut and then ran uphill as the tree slowly, slowly began to fall, finally settling with a whump into the duff and the slash. On these working trips, I had no fear of abandonment. Getting a load of wood took so much concentration that ditching a stepdaughter was a very low priority on Daddy's list. Even I knew that.

In the summers Daddy took us—sometimes all of us, sometimes just Elizabeth and me—into the endless maze of ridges and canyons, mostly belonging to the Beaver Creek drainage, where the logging was. We saw giant loaders parked on the landings; we saw the spar trees on the high-lead side, draped with cables, a jumble of logs at their feet waiting to be lifted onto the trucks on Monday morning. We saw the big D-8 and D-9 Cats parked like resting dinosaurs on the flatter logging sites, "the Cat side." We looked up and up at ponderosa pines and Douglas-firs, over 250 feet tall, and listened while the afternoon breezes grew to a roar in their branches, and the swaying of their tops turned us dizzy as we craned back our heads and watched.

We smelled, for the first time, the tang of thousands of growing conifers and the soft wet smells of the disturbed forest soil, the turpenes of oozing sap on fresh stumps. We looked out over just finished logging shows that seemed like the aftermath of a hurricane—the slash six feet deep on the ground and, standing amid the carnage, a few damaged "seed trees," as they were called in the tongue-in-cheek parlance of those wide-open logging days. Fruit Growers was just beginning the second round of timber high-grading on its own lands. The Douglas-firs and true firs and incense cedars left behind at the first go-around now fell in their turn, responding to the booming building business in California.

Fruit Growers' Cats dragged logs down the creeks—"Nature's skid trails," Daddy called them wryly. It damaged the creeks, he admitted, but it was the fastest way to get the logs out, and the Company could do what it wanted on its own lands. The Company didn't pile its logging slash, or burn it, because that cost money. On neighboring Klamath National Forest lands, slash was piled and the piles were burned, and Cats were supposed to stay out of the creeks.

Driving home in the twilight, past the double ranks of great trees, black against the darkening sky, I felt very small and humble, and useless. How could I—or Elizabeth, or even Mother—be of any importance to Daddy, when he could come here every day? We must seem less than nothing to him, compared with this. I wondered that he bothered to come back to us at all. Why would he want to live in a house among dry foothills, when all this waited for him, where the trees spoke to the sky? We lived the lives of ants, never seeing all this;

he lived a grander life, doing important and daring things, things that took skill and courage. He hiked alone through places roadless since time began and, with compass and clinometer to guide him, hung the colored flagging that would lead the way for the road builders and then the loggers. He watched the logs come out of the woods, and when he came back to Hilt at night he told us about the deer and bear and hawks he had seen.

In Beaver Creek's seventy-thousand-acre watershed, we saw the places where the fires of '55 had burned hot, and we saw where Dad hunted deer every year in the brush fields growing up around the black snags. Before the Haystack Burn, in fact, deer hunting was relatively poor around Hilt. A century of market hunting, followed by subsistence hunting in the 1930s, combined with fire suppression that reduced the number of sprouts and young brush that deer needed for food, kept populations low. Selective logging of ponderosa and sugar pines did not encourage the growth of the forage plants that deer preferred. But the Haystack Burn created thousands of acres of brush and sprouting hardwoods, and deer numbers exploded.

Ten thousand years ago, as the mountain glaciers of the Ice Age melted away, the first people of the Klamath Mountains found their way here, following the rivers and creeks, establishing villages along the river. The Shasta people had a village near the mouth of Beaver Creek. Before they came, at least one earlier people had lived there—people who used stone mortars and pestles. The Shastas, who did not use such tools, called them sacred and did not touch them.

The Shastas burned the areas around their villages every autumn, to cleanse the hills, encourage the growth of *icknish* (wild celery), and to discourage poison oak and drive away ticks. New browse sprouted to feed the deer and elk that wintered in the river canyon. Above the river in the canyons of its tributaries, they spread fire to encourage the new growth of bear grass, used for basketry, and to keep clean the groves of white and black oaks where the people gathered acorns.

Along the river itself, they established fishing stations for catching salmon with dip nets, or speared them as they ascended the creeks. The first salmon netted each fall were always released again to continue upriver, for they had been blessed by the first-salmon ceremonies of the Shasta tribe's downriver neighbors and cultural mentors, the Karuks.

The Beaver Creek drainage held over thirty miles of spawning grounds for coho and chinook salmon and steelhead. Within its fifty-four miles of tributaries lived rainbow trout, speckled dace, small-scaled suckers, marbled sculpin, and lampreys. The young of the steelhead lived in the creeks for three years, then journeyed downstream with the spring floods, to mature in the ocean two hundred miles away. The young of the lamprey burrowed into the streambeds, where they waited seven years before emerging to follow their steelhead prey, sucking blood with their round, hook-laden mouths.

High up in the drainage, in the cold seeps and springs and elk wallows, red-legged frogs and yellow-legged frogs woke in the mud to the slow-melting snows. In the swift cold streams near the Siskiyou Divide lived tailed frogs, strange, primitive creatures, hiding beneath rocks. In Jaynes Canyon and on the West Fork of Beaver Creek, in Trapper and Hungry and Grouse Creeks, the Pacific giant salamander laid its eggs.

Under the downed logs and rocks lived terrestrial salamanders, hiding from snakes and shrews. Within the rotting logs, fungi spread their mycelia out toward the young roots of pine and fir and twined around them and through them, giving them nourishment. The fungi grew and created fruiting bodies for voles to find and devour, spreading the nurturing spores throughout their burrows, which also gave oxygen to the soil. From high in the timber canopy, hung with lichen, spotted owls glided in the night, seizing rodents. With the morning, martens cruised the trees looking for squirrels, and fishers and wolves and wolverines passed through, silent under the old trees.

In summer, herds of Roosevelt elk traveled up to the high meadows under the Siskiyou Divide, and in autumn the Shastas drove them with a special breed of hunting dog—destined for oblivion, like their masters—as the white miners slaughtered the elk with guns. Black and grizzly bears ate fish and acorns in the fall and dined on newborn elk calves in spring, digging for grubs and roots in the wet meadows that the beaver had created all up and down the drainage that would someday be called Beaver Creek.

The beavers' impoundments, filling in over the centuries, built the meadows that fed the bear and the elk. They built the dams that made the quiet water where young salmon and trout hid and fed. Beginning in the 1820s, they were caught in steel traps set by the Hudson's Bay Company men, and by 1840, Beaver Creek and all the surrounding drainages were almost empty of beaver. A remnant population had barely begun to recolonize the drainage when, in 1850, another invasion began. This one would rip their dams apart, wash away their ancient meadows, and blast their universe aside. The beavers that escaped that destruction retreated to the river, where they hid in the riverbanks, and their children built no dams.

The world that I first saw from the seat of Daddy's pickup truck was a world that had begun unraveling long ago, beginning with the miners who destroyed the creek beds, drove the Shastas from their villages, and dug ditches to channel water to mines and blast away the spawning grounds of salmon and steelhead. The first miners moved on, and homesteaders and ranchers and still more miners came, to build bigger ditches for mining and irrigation ditches that diverted young salmon and spawners alike into pipes or out onto hay fields. The cattle of the homesteaders and ranchers grazed the riparian areas to dust and broke down the streambanks and widened the streams, warming them to temperatures lethal for young trout. Silt washed into the streams, and salmon eggs suffocated.

By the 1860s, only large mining companies had the money to build the flumes and ditches needed for the new hydraulic mines that could uncover the gold hidden in old river terraces high above the creeks. Several mining companies built a wagon road up Cottonwood Creek, over the saddle later called Four Corners, and down into upper Beaver Creek. Creek banks and inner gorges were blasted apart by torrents of water. The narrow floodplain of Beaver Creek was stripped once again of soil and vegetation. When the great Christmas Flood of 1964 scoured the watershed a hundred years later, it merely turned an already impoverished riparian world over one more time.

As timber was cut along the creeks, the banks caved in. In 1886, a man named Patterson hired twenty-five men to work on a placer mine in the main channel of Beaver Creek, above Hungry Creek, and the entire length of the stream ran dark with mud. On the Pacific Coast, at the mouth of the Klamath River, canneries processed salmon without end. Patterson's men bought the canned salmon, ate it, and threw the cans into the gullies they had made.

In the 1890s, as farmers in California's Central Valley grew tired of silt and debris from hydraulic mining in the Sierra Nevada clogging their ditches and fields, they pressured the state legislature to pass laws limiting the seasons of such mining. With the new laws, miners turned to hard-rock mining instead. On the ridges above Hilt, the Mount Bullion and Sterling Mill Mines opened, and with rising copper prices came the Siskiyou Copper Mine. By 1912, the Flystain and Deer Creek areas were producing gold, and shortly afterwards, the wagon road from Hilt was extended down Hungry Creek, up the Cinnabar Springs Road to the new Cinnabar Springs Resort and its hot springs, all the way to the Cinnabar Mine, which produced mercury. The Corbett Mine in the Bullion Mountain area hit a thirty-thousand-dollar strike, but none of the other seven mines nearby did as well. Hard-rock mines actually produced more profit for sawmills than for miners, as the demand for lumber for use as mine supports and stamp mills increased.

The 1890s also brought the first dredges to the creeks, and these operations continued for fifty years in Siskiyou County. Many hard-rock mines shut down during World War I, never to reopen. The rest closed down during World War II, except for a few war-related chromium operations like the Snowy Ridge and Starveout. They, too, were gone after the war.

Many people had moved to Beaver Creek and the Klamath River country during the Depression, thinking to wait out the hard times. They mined a little, hunted, fished, gardened, raised a little livestock, and poached. They claimed to be miners, but earlier prospectors had been thorough, and there was little left. The Forest Service knew that most of the people living on the public lands weren't really miners, but local rangers told Washington to cool it, that if they tried to evict these people, there would be violence. Let them stay until times were better, they advised.

Over its lifetime, the Fruit Growers' sawmill at Hilt milled about a billion and a half board feet of timber. Much of it came from the Beaver Creek drainage. By the early 1950s, it was possible to circumnavigate that entire drainage by road. The logging and road building that followed the Haystack Burn were only an acceleration of something already underway and begun long ago. Its consequences continue today.

The eastern half of the drainage is part of the Mount Ashland pluton, a massive granitic intrusion slowly crumbling as erosion exposes it. The western half is metamorphic and includes the Condrey Mountain schists, once crushed under enormous geologic pressures into a uniform bedrock, now exposed and weathered into beds of "blue goo" on the Siskiyou Crest.

The crest, from Condrey Mountain on the west to Mount Ashland on the east, forms the northern boundary of the drainage. Ten thousand years ago, glaciers scraped the crest bare. The thin layers of soil that managed to form with the return of warmth and vegetation were destroyed when uncontrolled livestock grazing, from the 1860s to the 1930s, stripped the land again. The western boundary drifts down from Condrey Mountain, south along a ridge to Dry Lake Lookout and Round Mountain, and finally down the brushy divide between Fish Gulch and Quigley's Cove Gulch to the Klamath River. East of the mouth of Beaver Creek, the ridges climb north again, to Buckhorn Bally Lookout, to Bullion Mountain, across a big saddle to Hungry Creek Lookout, over to the Four Corners saddle, and then straight up north to Mount Ashland, which, at a little over 7,500 feet, is the highest point on the watershed's boundaries.

At peaks like Dry Lake and Buckhorn Bally and Hungry Creek, the Forest Service had maintained lookout towers since the 1930s. And it was from the catwalk of the Buckhorn Bally Lookout that I first saw the immense size of the Klamath Mountains. From its 5,100-foot elevation, we saw not only the surrounding peaks and drainages, and the ethereal double volcanic cone of Mount Shasta, fifty miles away, but also fourteen million years of peneplained landscape. We looked downstream to the mouth of Beaver Creek, where groves of oak baked in the hundred-degree summer heat, and upstream to the big alpine meadows below the ridge called Dry Lake, where a few patches of snow survived all summer. The mouth of the creek received twenty-four inches of rain in a year; Dry Lake absorbed sixty inches, mostly as snow. The wet season ended abruptly in late May; there was no more rain until late October or early November. Every few weeks in summer, lightning storms built up over the mountains south of the river, marching northeast, bringing, at last, the fires of autumn.

Carl and Matty, the Buckhorn Lookout guards, were retired Canadians, who in winter lived in a house on the Klamath River. Carl was short, with a shock of thick white hair and an endless fund of stories. He was, he told me, a great-great-nephew of Tecumseh, which I later discovered placed both of our ancestors at the abortive Battle of Point Pleasant. Matty was even shorter than her husband, with cropped iron-gray hair and the darkest tan I had ever seen. She was addicted to nude sunbathing, she confessed to Mother. When we visited them, Daddy's pickup groaning up the steep, spiraling road that ended abruptly at the foot of the lookout tower, we stepped out into a high, dry, eerily quiet world, isolated and utterly captivating.

Carl and Matty fed the wildlife, so here, chipmunks ran right up to our feet, and a doe snorted at us from the salt block just below the tower, annoyed at seeing strangers. In the mornings, they told us, a hen grouse led her family to drink at the spring that trickled out from the rocks on the north side. Inside the lookout, there was a 360-degree view through curtains of glass. A double bed stood on four thick glass feet, as did a couple of short stools. During a thunderstorm, Carl and Matty couldn't answer the telephone, and all communication was by radio, while standing on one of those glass-footed stools. Carl showed us how he used the range finder mounted in the center of the floor to locate a plume of smoke before radioing its legal location—township, range, section—to the Forest Service dispatcher in Yreka. Periodically, while talking to us, one of them picked up a monstrous pair of binoculars and scanned the horizon.

Our family outings with Daddy always produced some sort of haul: blackcap raspberries and elderberries for jelly; elderflowers and pipsissewa root for tea; jugs of mineral water from Cinnabar Springs for lemonade; rocks for a garden wall; river silt for the garden; fish, wood, deer. Daddy brought us into his family's hunting and gathering tradition. Daddy's example taught us that the natural world—the forests, the rivers, the creeks, the ocean—was not a playground. They were places to search for and gather useful things. It was not so important that the things always be found; indeed, nine out of ten deer hunting trips were failures, in the sense that no deer died. But deer were a high-energy quarry. Low-energy quarry—like berries, which you could always find if you went looking for them at the proper time—was a sure thing. But what was somehow important, we came to understand, was the search itself. To this day, I have a hard time going anywhere in the outdoors just to have fun. Rock climbing or downhill skiing—something deep within me tells me—are not activities upon which the land looks kindly. My sister says it best, perhaps. "We didn't go on vacations," she remembered not long ago. "We went on food-gathering expeditions."

CHAPTER 16

> If children were brought into the world by an act of pure reason alone, would the human race continue to exist?
> ARTHUR SCHOPENHAUER

The prosperity of the late 1950s increasingly brought things *to Hilt, new* things. More and more, the people of Hilt worshiped at the altars of leisure and entertainment and possessions. Shirley DeClerck went to work at the Company office, even though her husband, Audomar, had a perfectly good job. They bought horses and went riding on weekends with the new resident manager's family. "Social climbers," Mother sniffed.

"People want to be entertained all the time now," Daddy said to a car dealer who wanted to put a radio in the Ford sedan. "Nobody knows how to be bored." Elizabeth and I thought the world contained quite as much boredom as was good for it.

The two of us couldn't figure out why, if Daddy enjoyed watching the Three Stooges on television (more than we did, in fact), he didn't want a television himself. He shook his head at our Hilt neighbors who bought outboard motorboats, then built carports to shelter them. He refused to buy us puppies or parakeets, and even though Carleen and Barbara DeClerck, and Marsha Cavin, and Mimi Barron now had horses, he positively refused to invest in a pony. So it came as something of a shock when Mother told us that she was going to have a baby.

Some months before our little brother Tommy was born in May of 1959, Mother told me The Facts of Life. I had, of course, seen bulls covering cows in pastures, and I knew that cows had calves. And I had watched Blackie, an old cat who for years ate table scraps off a table outside Grandmother's back door, give birth to three kittens, lying in a pool of birth fluids and sunlight on the front porch.

Pregnant women were scarcely unknown in Hilt, either, and I knew there were babies in those big stomachs. But as Mother, making cookies in the kitchen, began to explain to me how the babies got *into* the women's distended bellies, I gaped at the lengths to which grown-ups would go to find disagreeable ways of spending their time. I heard with skepticism the news that sex—always between married people, of course—was an expression of love. This was the same woman who said she was giving us cod liver oil and taking us to the dentist because she loved us.

Thanks to Robbie, Gino's art collection, and Mother's old copy of Gray's *Anatomy,* I had a working knowledge of the human organism. But this mechanical joining of men and women seemed inexplicable, with no connection to the real world.

With a younger sibling imminent, I pondered another mystery. Why did our parents have us, in the first place? Once I understood that parents actually *did* something that created children, the logical extension was that grown-ups ended up with children because they wanted them. I found this astonishing. Once, I had thought that parenthood just happened, like getting your big teeth. But if our parents deliberately created us, why didn't they seem happier to have us around? Why did Daddy complain about how much my glasses cost?

If children were a burden imposed by a malign fate, it made sense that parents complained about glasses and outgrown shoes. Santa Claus made sense as well, for why would our parents, who were never happy with us, give us Christmas presents? We ourselves knew how much fun it was to torment the less fortunate—to laugh at Alfred because he was fat and slow and flatulent, and at Linda when she simultaneously caught ringworm and poison oak. So our parents must punish us because they enjoyed it. Having lost the ability to have fun themselves—for they no longer wanted to go sledding, or ride bikes, or play baseball—their only source of enjoyment must be tormenting us, their unwanted burdens.

But if our parents produced us after due consideration, then the world of punishments and rules made no sense at all.

One day when Mother and Daddy were discussing a childless elderly couple, Daddy said that it was sad that they had no children to take care of them in their old age, and the light dawned. I could see where this one was going. Daddy and Mother put up with us because they expected us to come back and wait on them when they were old. That was why people had children: for security, not because they really liked us. The world fell back into place. The new baby was extra insurance in case Elizabeth and I didn't work out.

If we thought that Mother's temperament was unpredictable before her pregnancy, we had not seen the half of it. As she grew larger and more unwieldy, she grew more cross, and so did Daddy. In the evenings, Daddy escaped to his workshop, where he was building a cradle for the infant. In the afternoons, Mother put on a scarf and the big coat she couldn't button anymore and performed her daily walk, half a mile out on the county road toward the SS Bar Ranch and back. Elizabeth and I cruised around her on our bicycles. On the last day of May, following two false alarms, our little brother was born in the hospital in Medford.

Thomas Lee Brannon was the smallest human being I had ever seen, less than seven pounds at birth. As we peered through the hospital's nursery window at him, I couldn't believe it. His bald head was the size of a small grapefruit. The mountain that was our mother's stomach had produced a small naked rodent. He was red and his eyes were screwed shut as though determined not to like this strange new place. In her room, Mother looked pale and tired and her abdomen still looked enormous, as though someone else might lurking in there.

When Tommy came home, his life—and ours—was suddenly taken up with the endless round of his bodily functions, all of which were either smelly or noisy. Unattractive as he was, we nevertheless begged to hold him, and we marveled at the delicacy of his translucent little fingers. We touched the soft spot on top of his head, smoothing the white, downy hair. We watched while Mother changed his diapers, and we learned how to do it ourselves, our fingers carefully shielding his skin from the sharp points of the safety pins.

Tudy arrived in Hilt shortly after Tommy—her mission: to help Mother and make our lives a living hell. Her idea of helping Mother was to take over the baby's care so Mother could go back to housework, but somehow she always managed to be doing something else when Tommy's diapers needed changing. She ordered Elizabeth and me around, argued with Mother about the proper treatment of Tommy's umbilical scar and circumcision, and commandeered the house so effectively that Grandmother stopped coming to visit entirely. For us, evenings were the worst. Daddy could hide out in his workshop, but Mother and Elizabeth and I had to sit up with Tudy and listen to her endless stories, to a background of wailing baby. We stayed overnight at Grandmother's house as often as we could get away with it.

The only good part of Tudy's visit was that Mother, constitutionally incapable of criticizing her mother-in-law to her husband's face, was forced to complain to her daughters, and all at once, the easy, sisterly comradeship we had almost forgotten came back for a while, to be reawakened in the years to come every time Tudy came to visit. Suddenly, there was a subject about which we were all in agreement.

The longer Tudy stayed, the more interest she took in our upbringing. We heard her complaining to Daddy that she had seen Some Nuns getting out of a car down by the church. Yes, of course you did, he told her. They come every week to give catechism classes to the Catholic kids. Tudy's chest expanded with indignation. The Girls Don't Go, Do They? she demanded in capital letters, fixing her slightly protuberant eyes on us. We shook our heads, but we were, of course, lying. We *had* gone to catechism with Nancy Trinca only the week before.

We had felt like intruders as we walked into the familiar little church in the middle of the afternoon. Two nuns, a pretty young one and a sturdy older one with a hint of gray mustache, both in full pre-Vatican II penguin regalia, walked back and forth between the church and their station wagon, carrying books and calling out things like, "We're almost out of holy water, did you remember to bring it?"

We copied Nancy's movements as she dipped her fingers in the little dish on the wall inside the door and crossed herself. We sat down with her in one of the front pews, noticing that all the kids bobbed a curtsy before entering the pew, just like we did at St. Mark's Episcopal in Yreka. We watched as the sisters swished up and down the aisle, laying out their teaching materials. I stared at their full-dress uniforms of floor-length black habit and white wimple covered with a black veil. The younger one smiled at Elizabeth and me, the newcomers, and asked, "And are you Catholic girls, too?"

"Anglo-Catholic, yes," we replied, well coached by our rather High Church mother. The two black veils seemed to shudder a little, but they nodded and let us stay. As I listened to the other kids recite, it seemed to me that there wasn't that much difference between the Catholic catechism and the Episcopal one in which Mother intermittently drilled us. The narrative lesson for the day was about church history, specifically the part about Henry VIII of England and his marital troubles. It must have seemed too good an opportunity for the older sister to pass up, although she interrupted herself several times to hope that none of this was offensive to us. We answered brightly and quite truthfully that, no, it was very interesting. And it was.

Since Mother's marriage, our catechism lessons had become infrequent. We so seldom went to services at St. Mark's that Mother must have felt that the likelihood of our being able to attend confirmation classes any time soon was remote. The community church in Hilt was shared by Roman Catholics and Protestants on Sundays—a generic Protestant service in the morning, a Catholic mass in the afternoon.

Besides reading Grandma Rosa's Christian Science tracts, we often discussed points of theology with Mother, who was a sincere and faithful Anglican. Daddy continued to regard churches as swindles and enemies of common sense and logic. So it was very annoying when, on the rare occasions when we went to church, Daddy listened with attention to the sermon and always discovered some sort of moral lesson directly applicable to our misspent lives. Elizabeth and I were purely liturgical Christians; we believed in the magic of the sacraments and the candles and the incense, in the hair-raising skirl of the organ and the sun shining through the Agnus Dei window above the baptismal font. Daddy believed most people would be better off if they went out into the woods on a Sunday, instead of sitting in a church or reading the Bible. "You can learn more about a painter by studying his paintings," he said once, "than you can by reading his biography."

We had enough sense not to tell Tudy that we had already sat at the feet of the Whore of Babylon and to hide Grandma Rosa's gaudy books away in the bottom of our toy chest. But her visit held doctrinal pitfalls we had not counted on. One day Tudy found me reading an encyclopedia of natural history that Jo and Carl had sent me for my birthday. I especially loved its lengthy descriptions and imaginative illustrations of prehistoric animals. When Tudy asked me what I was reading, I launched into an enthusiastic explanation of evolution and showed her the pictures, including one of a hairy Australopithecine morphing through several incarnations into a Cro-Magnon, who bore a striking resemblance to Richard Widmark. Tudy leaned forward, reached out her large brown hand, and slapped me so hard that for a few moments I was afraid that the silver Navajo bracelet on her wrist had broken my jaw. We were alone in the house; for once Mother had been able to put Tommy in his stroller and take him down to Grandmother's for a visit. I remember backing toward the front door and somehow getting through it, the maddened blue eyes following me.

When Daddy came home that evening, Tudy went out to meet him at the back gate. Elizabeth and I raced out the front door, down the steps, through the rock garden on the south side of the house, and flattened ourselves against the wall of the workshop, next to the alley. We tried to overhear their conversation. Tudy was remonstrating with him, and we caught a few of her low and urgent phrases: "What are you teaching those girls?" floated across to us. We were delighted when Daddy—tired, sweaty, his clothes streaked with soot from the little lightning fire he and a crew of loggers had found and put out that morning—retorted sharply that he didn't believe in Adam and Eve either, and he wasn't going to be a hypocrite and tell us that he did.

At supper that night, Tudy was unusually quiet. Mother looked from Daddy to Tudy to the red mark on my face with puzzled displeasure. The next day, Daddy was cross and impatient with his mother, Mother was almost but not quite rude to her mother-in-law, and the day after that we all watched with undisguised relief as the big yellow Pontiac station wagon drove away.

CHAPTER 17

> We used no mattress on our hands,
> No cage upon our face;
> We stood right up and caught the ball
> With courage and with grace.
> GEORGE ELLARD

I cannot remember a time when I did not know how to hit a ball with a bat. Behind Grandmother's house, across the alley, was a vacant lot we called "the playground." Long before I was born, it had boasted a tennis court of packed clay, surrounded by a high mesh fence. By the time I saw it, all that remained was a round chunk of concrete, battered and pounded and stuck in the earth about ten feet from the back wall of the garage building that was the playground's western boundary. The garage wall made a good backstop, and the chunk of concrete was perfect for home plate. Sometimes a ball got by the catcher and rolled through a hole in the wall, and then someone had to run to the end of the long, unpainted building and brave the dark and creepy crawlspace to retrieve it. The old boards were infested with black widow spiders, and scorpions lurked beneath the rocks on the damp ground.

If we had only three or four kids available, we played Pinkie, with one batter. If a fielder caught a hit ball in the air, he traded places with the batter. If the batter hit a grounder, he tried to run to first base and then back home before being tagged or thrown out. The very youngest kids learned to play ball this way, with the pitcher carefully moving up a few paces and lobbing the softest of underhand pitches toward the plate. If any contact at all was made between ball and bat, we all yelled at the little person to drop the bat and run to first base, often having to point out the square of cardboard lying in the depression at the end of the hard-packed path that was the first base line.

With enough kids on the field, we played workups. Batters thrown or tagged out went to the outfield while everyone else moved up one position. If a fielder caught a ball on the fly, they could either trade places with the batter, or magnanimously "give it to the game." In the fall and spring at school, we played workups endlessly during the long noon and afternoon recesses. Extra players were stashed in the grass-grown outfield. We seldom had eighteen players at one time except during official PE periods, so regulation games were rare. By the time I was eleven, a new Little League field had been built just across the railroad tracks from the Company store, but girls weren't allowed to play Little League then, and there was no girls' softball league. I went to watch the games sometimes, holding my Roberto Clemente glove and slapping a hardball into it, inhaling the fumes of leather and neats-foot oil wafting up in the heat.

We had never heard of the infield fly rule, and when it was pointed out to us at school, we thought the teacher had made it up. We played with any ball we happened to have—a baseball, a softball, a discarded and hairless tennis ball. We had our own rules, like One Base on an Overthrow (necessary because most of our outfielders couldn't throw and most of our basemen couldn't catch), and any ball hit past the swings and the teeter-totter into the high weeds in the outfield was a double, as well as any hit that entailed climbing a fence.

The early 1950s were the end times for community baseball in Siskiyou County. Hilt had a long history of baseball teams sponsored by Fruit Growers and before that by the Northern California Lumber Company. Before 1920, teams from Hilt played as far away as Portland and Sacramento. In the 1920s, a Hilt ballplayer named Ken Williams played for the St. Louis Browns. After World War II, Raymond Coleman, Orson's son, played professional baseball for seventeen years, and about 1950 a big new baseball diamond, with fences and bleachers and dugouts, sprang up overnight after the Company leveled a few acres of sagebrush on the south side of town. The local town league included Yreka, Montague, and Fort Jones. Some reservoir of optimism seemed to inhabit Hilt's baseball teams right after the war. Perhaps it was simply joy in being alive, in looking forward to a time of peace and plenty and a lumber market with no crest in sight. But the companies who sponsored such town teams were not slow to notice that the GI Bill had provided them with a reservoir of college men who worked only in the summer, and some people said that was what killed community baseball, that the local boys back from the war couldn't compete with the college ringers.

High water mark for the Hilt Jets, as the town team was known, came in 1952, as a team that included Gino Michelon, Gino Trinca, Tony Marin, and Robert Trinca won the county baseball championship. I remember going to at least one of their games and, from high up in the covered bleachers, watching through the wire mesh behind the plate, hearing the bats crack, and smelling the dust and leather as baseball was played as it ought to be—in the dirt and the wind and the sun.

Semi-pro ball in Hilt was killed not by ringers or corporate competition but by a choice of place, of where to be on a weekend, made possible by the automobile and unrationed gasoline. No one had to stay in Hilt on a weekend anymore. A baseball game was not the only entertainment available. The young men courted and married, or moved away, and wives and children intervened in their lives, and all the distractions of a larger world fell open before them.

I think we would have waited a year or two longer for television if CBS hadn't carried *The Baseball Game of the Week.* For that, Grandfather at last broke down and invested in a television set in the summer of 1954. On Saturday mornings, if we weren't going fishing or to town, the set went on, and we learned to worship new gods.

Grandfather liked the Yankees and the Giants. He must have been the only Yankee fan in Hilt who admitted it. In those days, CBS seemed to carry a lot of American League games, so I saw a good deal of the Yankees, but my favorite player was not the golden boy, Mickey Mantle, but Yogi Berra, the clutch-hitting catcher built like a fireplug. Yogi seemed to prove that you could he homely and still be competent and beloved.

I collected baseball cards by gathering pop bottles off the roadsides and turning them in for the deposits, until I had enough money to buy a pack of cards. Elizabeth and I traded cards at school, and I managed to amass a respectable Yankee collection, since many of the other kids got rid of them as soon as possible. I could have all the Tom Treshes and Tony Kubeks I wanted.

I read the *San Francisco Chronicle's* green sheet and followed the standings and sat beside Grandfather on the couch, as he showed me how to read a box score. Between us, we had seen exactly one major league baseball game in our lives: White Sox vs. Tigers, Comiskey Park, August 1956, during a two-week pilgrimage hack to Iowa to visit Grandmother's relatives. The Tigers won. Grandfather had, however, seen several Triple-A league teams play, including the San Francisco Seals, who never quite got over the glory of having once nurtured Joe DiMaggio.

In 1954, the St. Louis Browns moved to Baltimore, and the team's terrible luck began to change. By 1960 they were ready to overtake the White Sox for second place in the American League and give the Yankees some competition. I began to notice that they beat the Yankees fairly regularly in hard-fought, extra-inning contests, but the Yankees themselves were not what they had been. In that year, they had only one .300 hitter and scarcely a pitcher who had worked two hundred innings. Roger Maris brought his bat from the Athletics, but Mantle was plagued by injuries.

When the Pirates won the National League pennant, Grandfather grunted. Sure, Bob Friend and Vern Law and Roy Face were adequate pitchers, but their only really threatening hitter was Dick Groat. Still, the Pirates won the first game, 6–4, giving their fans hope, although they then proceeded to lose the next two games in spectacular fashion, 3–16 and 0–10. They managed to eke out wins in the fourth and fifth games, 3–2 and 5–2, before collapsing utterly in the sixth, 0–12. I asked Grandfather if he thought the Pirates had any chance at all in the seventh game. None, he assured me. And this must be true, I thought, because Grandfather always knew what he was talking about when it came to baseball. It was, he had even been heard to say, in his blood.

Edd Roush, Grandfather's fourth cousin three times removed, won the National League batting title in 1917 with an average of .341. He played in the 1919 World Series and always said that his Reds would have beaten the White Sox even if the gamblers hadn't gotten to them. He played for Cincinnati for eleven years, from 1916 to 1926, coming there from the Giants along with Christy Mathewson and Bill McKechnie. Roush began playing baseball in his native Indiana, then played for a Chicago White Sox farm team in Lincoln, Nebraska. He played for Indianapolis in the short-lived Federal League, going to the Giants when the Federals folded in 1915. He hated John McGraw of the Giants, who returned the favor and threw him in with the trade for Mathewson and McKechnie.

Roush was fast; he stole nearly three hundred bases one year. He led the National League in batting twice, and hit more than .350 for three years in a row before slumping all the way to .348 in 1924. The lowest he ever hit in the majors was .321 in 1919, and that was still good enough to win the National League batting title that year. After 1926, he played three more years for the Giants, where his three-year, seventy-thousand-dollar contract made headlines. He was now in the Baseball Hall of Fame, but none of the other kids in Hilt had ever heard of him. I was proud of the connection, but it was like having a president in the family that no one had ever heard of, either: Harrison, or Fillmore, maybe. After a while, you just didn't mention it.

In those days there was only the World Series. And it was played in the daytime, with no concessions to television. If you were at work or at school when the games were played, tough luck. So the final game of the 1960 World Series, Yankees vs. Pirates, came to pass on a weekday afternoon. Someone had brought a television set into the school and laboriously hooked it up to an antenna during lunch hour. Perhaps this was simply a way for the teachers to keep track of the game's progress during their lunch hour, but the inevitable happened—after lunch, no one could bear to turn the thing off. Gradually, all of the fifth through eighth grades filtered into the seventh and eighth grades' room, and no one chased us away as we watched the final innings.

I remained convinced that the Yankees would win, even as the score shifted back and forth, inning by inning: 2–0 Pirates, 4–0 Pirates, 4–1 Pirates, 5–4 Yankees, 9–7 Pirates, and finally 9–9 as the bottom of the ninth began. I felt no qualms as Bill Mazeroski stepped to the plate to face Ralph Terry. He had hit only eleven home runs in the regular season, and Yankee pitching was surely good enough to demolish him and the rest of the Pirates batting order.

Ralph Terry pitched, and Mazeroski swung and hit the ball. He hit it over the infield, and over the heads of Boyer and Mantle and Berra, and over the left-field wall, where it disappeared into the crowd. All around me, kids erupted out of their seats, screaming with joy, jumping up and down on top of their chairs, hugging one another.

I sat frozen, unbelieving, aghast. Delores Luper danced by me, a look of demonic triumph on her face. She leaned over the desk where I sat, her big horsy teeth bared in a sneer. "Nyaah! Nyaah!" she shrieked at me, her arms akimbo. "Your Yankees aren't so big *now,* are they?"

My Yankees. She made it sound like an obscenity. I stood up, banging my knee on the hard underside of the desk, the pain and emotional reaction making tears start from my eyes. I screamed back at her, "The Yankees are the greatest team that ever was!" so loudly that the spit caught in the back of my throat and I started coughing. She laughed and stuck out her tongue at me and danced away, bouncing in victory.

I could not believe it then, nor later as I walked home. The Yankees had lost the World Series, even as they set Series records for most runs, most hits, and highest team batting average. For the next few days, every time I opened the sports section of the newspaper, I hoped that the headlines would read, "Terrible Mistake Made: Yanks Win After All." The Yankee management evidently felt the same way, for they fired Casey Stengel. To me, it was as though the angels had fired God.

The joy that broke out in that classroom as the Pirates won puzzled me for a long time. It would be many years before I fully understood that behind that celebration stretched a long history of Us vs. Them, Workers vs. Bosses, Front Street that always won and Adobe Street that always lost. Delores Luper and I were on opposite sides in that long war, but we had not chosen our sides; they had been ordained long before we were born.

Grandfather believed that if you worked hard and kept your nose clean and didn't do stupid things, life would reward you; his entire life had been built on that belief. Bill Roush didn't need anybody's help. Bill Roush didn't need a union to take care of him. He had made it on his own, and people who couldn't make it on their own were weak. He saw in unions a haven for whiners and slackers. Although he had been named by his staunchly Democratic father for William Jennings Bryan, the Great Commoner, Grandfather was a Republican in a town that had voted Democratic in every election since 1920.

Most of the plant foremen, remembering their origins, or perhaps just more aware of how many unfortunate accidents can happen in a sawmill, at least tried to be one of the boys. Grandfather never did. In front of his employees, he identified completely with the Company's interests and made it clear that he regarded unions as the enemies of initiative and free enterprise.

Fred Haynes was a machinist and the Hilt representative for the Lumber Industry Union. He was active in community affairs, organizing a band (he played the trumpet), and joining the volunteer fire department. Fred showed up at the box factory in his role as shop steward just often enough to annoy Grandfather, who was convinced that Fred was little more than a union spy. They grew to hate each other.

Fred's son, J. Fred, went to work in the box factory in 1938. Grandfather figured that J. Fred was repeating everything that happened in the factory to his father and called him a complainer. J. Fred put up with Grandfather's abuse for two years, then escaped to the Company office.

After J. Fred left, the older Haynes still, in Grandfather's opinion, knew too much about what went on in the box factory. Grandfather suspected everyone, but took most of his anger out on Mr. Gilbert, the underforeman, who suffered silently, and for years, under Grandfather's loud, aggressive abuse. Many years later, Mr. Gilbert's son Billy cut Mother dead at one of the few Hilt Reunion Picnics she ever attended. "I felt bad about it," she said, "but I understood. As far as Billy was concerned, I was cut from the same cloth as Grandfather, and that made me just as bad as he was."

"Why do the Democrats always win here?" I asked Mother one November day, still in shock that Kennedy had actually won. First the Yankees, and now this.

"Because this is a working man's town," she said, with the mixture of resignation and bitterness that always colored her voice when she discussed Things That Couldn't Be Helped. The election results looked very different to most of the millworkers, of course. But no matter who won the presidency, the Company, and hence a bunch of reactionary fruit ranchers, was still in control of their working lives.

Perhaps, as she introduced me to the political realities of Hilt that day, Mother was remembering The Strike. In the autumn of 1945, with the war over, the millworkers union called a strike at Hilt. Living costs were rising with the end of rationing, and there were shortages. The union wanted a raise. The Company didn't want to give them one. For several months, the mill was shut down, and no one worked.

Mother came home from her bank job in Ashland one weekend during the strike. She walked down to the post office to mail a letter, and as she stepped up on the Company store's porch, she saw a group of union men lounging against the railings near the pay phone outside the post office door. Most of them were strangers to her, but among them was Fred Haynes. She saw Haynes turn to the other men and say something, and she watched as the other heads swiveled to look at her. One of them laughed, quietly. The others just stared at her, hatred in their eyes. Front Street bitch. That bastard Roush's daughter.

Mother jammed her hands into her coat pockets and kept walking. She looked at Fred. "Hello, Mr. Haynes," she said.

He blinked back at her, surprised. "Hello," he said.

Mother went quickly into the post office to mail her letter. When she came back out, steeling herself against hostile eyes and whispered insults, the pay phone was ringing, and the men turned toward the booth, listening as Fred talked into the receiver. Mother walked straight down the steps and across the road and down the long exposed sidewalk, and as she climbed the worn wooden steps of home, she felt her knees shaking.

CHAPTER 18

Bird of the bitter bright grey golden morn . . .
ALGERNON CHARLES SWINBURNE

Tommy had just passed his first birthday when Daddy called us inside one evening as he came home from work. He set his black lunchpail on the table, flipped open the latches, and slowly turned back the lid. Inside were a pair of black shoe-button eyes, a black beak wide open with hunger and indignation, and a couple of tentative, electric blue feathers, lost in a mound of gray fuzz. The young Steller's jay, fallen from the nest, squawked, and loudly. And so we met Perky.

We ran over to Nana's and borrowed an old bird cage; we found an eye dropper and opened a jar of Tommy's strained liver, and after several hours of chaos, Perky sat snuggled in a nest of torn newspaper, eyes shut. He was the first real pet Elizabeth and I ever had, and from the beginning we knew he wasn't ours to keep. "You'll have to let him go when he gets old enough to fly," Daddy told us. But of course, by that time, he was used to us, and even after he was free of the cage, and roosted high in the locust tree at night, he still came fluttering down to the front porch in the mornings, his "wAAAK!" alerting us all. He begged for food, long after he could pick up ants on his own.

There never was a day that I was not afraid for Perky. The odds in Hilt did not favor young birds. There were boys with BB guns; there were cats; there were cars. But for enough time to let us become too fond of him, Perky led a charmed life.

That summer, Aunt Jo and Uncle Carl were living in an apartment in Ashland. Carl was out of the Air Force and working as the night chef at the Medford Country Club. Jo was pregnant, the baby due in October, and she was lonely in the daytime, so I was sent to stay with her for a couple of weeks. We went shopping at Woolworth's together, and to the movies, and on Sunday to services at a Presbyterian Church, because it was within walking distance. We sat through hellfire-and-brimstone sermons, while Jo inhaled her smelling salts and rolled her eyes.

I read all of Genesis and Exodus from Jo's white leather-bound Bible, enthralled by the narratives. In the evenings Jo and I watched television and ate popcorn and discussed babies. In the afternoons, I worried about Perky, and wrote letters to Elizabeth asking her how he was. I put the letters in the mailbox at the foot of the driveway and watched the mailman open the box and remove my letters and shove them into his capacious hag.

I picked up the little apples that fell off the tree outside the back door and bowled them down the wheel ruts of the driveway, so that they bounced over the curb and out onto the four-lane highway, to be squished by cars. I wrote more letters, asking how Tommy was, amazed to realize that I actually missed him.

While I was gone, our house in Hilt was painted. The Company, in the throes of some visiting executive's notion that the town looked "run-down," had pushed through a plan to paint all the houses in town. The idea that Hilt's wood-frame cottages should—almost forty years after their construction—be painted struck Daddy as a remarkably bad idea. "That dry wood will soak up a lot of paint; it'll take at least three coats before the job looks decent, and they won't want to buy that much paint."

He was right. Only the Front Street houses received three coats, plus a contrasting color for the trim. The rest received a single spray-painted layer, on body and trim alike. Beneath the thin patina of paint, the old wood soon showed through.

When the Company painters approached Daddy with color cards and asked which color he preferred, he told them that he preferred that it not be painted at all. Just paint the window trim white again, and it'll look nice and rustic, he told them. But they only stared at him, and under the silent reproach of their we-just-work-here-Mr.-Brannon gaze, he finally pointed to a color sample.

So I came home to find the house painted a startling light blue, and not until I saw the quirk around Daddy's mouth as we drove down into Hilt from the highway did I realize the joke. Our house was perched on a little rise at the south edge of town. Driving down into town from the highway a mile above, our house was, for every traveler, the first and most visible glimpse of Hilt. And it now looked exactly like a giant robin's egg with a porch.

Perky was hopping around the garden as we drove up, and as I approached he flew up onto my shoulder. His beak was blue. He had had a wonderful time when the painters were there, landing on their white overalls and pecking at their brass buttons and suspender hooks. And he had, of course, landed on the rims of the open paint cans and tried to drink the paint. "Well, he only did it once," Elizabeth consoled me. "He's into everything these days."

Perky had real feathers now, and there was no doubt that he could fly. More and more, he flew around town, extending his territory. He rode on Liz's shoulder up to Marin's one day and flew into the kitchen, lit on Tony's coffee cup, and took a swig. On his own expeditions, he tried to follow someone as they opened a screen door, and the door slammed on his beak, stunning him, and throwing the beak out of alignment. He startled housewives on their way to the store, as he landed in their hair, squawking.

I had been gone for two weeks, and everyone had survived without me, even Perky, a bird so trusting that his life was one long narrow escape. I spent the last nights of summer sleeping out on the lawn in my sleeping bag, and in the early morning, high above me in the locust tree, I heard Perky singing his imitations of robins and orioles, in the high sweet notes seldom heard from jays.

He began picking up pieces of colored glass in the alleys and tucking them into the holes in the pecky cedarwood fence between our yard and Johnson's. He was careless enough to roost on that fence in the afternoons, and twice Ruthie saved him from a cat about to pounce. And one day, of course, he disappeared.

We did not find him until October, when the little that was left of him turned up in the dry bed of Hilt Creek, recognizable at once because there were no other Steller's jays in town—they were a forest bird—and because of the distinctive beak. We should have taken him into the woods ourselves, and picnicked somewhere where the Steller's jays were, and left him there, I thought. A Bible verse, read that summer, ran through my mind: "Let me not see the death of the child."

On my way home from school the next day, I carried a handkerchief, and walking back up the creek bed, I scraped up the remains of beak and claw and feather with a wood chip. Walking back toward the alley, the light dry burden clutched in my hand, I was ambushed by Mike Trinca, who leapt out from the piles of old tires behind his Uncle Bob's service station and unleashed a barrage of crab apples at me. I ran, fleeing the missiles, ashamed of my cowardice, knowing that if Daddy learned of it, he would tell me what he had told me last year, when the boys threw clods of dirt at me.

"They'll keep throwing things at you until you beat one of them up," he had said. "If you don't have enough courage to do that, don't come crying to us." I was afraid of an actual physical conflict, and it made me ashamed. I said no more about the incidents, which continued. But this was too much. Mike, I was convinced, had killed my blue jay, knowing that Perky was mine, and had thrown the apples at me just to taunt me about it.

The next morning at school, I tapped Mike on the shoulder as he was hanging up his jacket, and when he turned around, I punched him as hard as I could in the mouth and kept on punching. Mike staggered backwards into the coats, tripping over someone's rubber boots. But although he was a year my junior, Mike was as tall as I was and heavier. He would never have punched a girl unprovoked, but after a moment's wild puzzlement, he evidently decided he was dealing with an unprecedented situation and began to fight back, fists whirling as he landed a hard punch on my ear that slammed the ear piece of my glasses into my skull. I grabbed his shirt and yanked and swung my fist at his eye socket, jamming my knuckles in the process. He fell on top of me, landing punches on my ribs, and I felt my head slam onto the linoleum tiles just as Mrs. Jeter grabbed Mike by his shirt collar and hauled him upright.

Sitting in the principal's office, watching the stunned look on Mr. Rhodes's face as he slowly pried the story out of me, I felt strangely peaceful. I didn't think I would be punished for hitting Mike, but if I was, I didn't care. Mr. Rhodes kept me sitting across from his desk until noon, then sent me home. I had a torn dress, a bruised face, and a new reputation. No one ever threw anything at me again. When Mother told Daddy, he only nodded. "The boy asked for it," was all he said.

That evening, Daddy dug a grave for Perky in the nasturtium bed and dropped beak, feathers, bones, and handkerchief inside, tamping everything down with a big round piece of river-washed granite and leaving it there as a headstone.

A few days later, I heard another boy bragging that he had shot a bluejay in town that summer and I knew that I had found the real culprit. I had never told David about Perky, precisely because I knew he had a BB gun, and I feared the generalized meanness of boys.

That fall, as baseball season gave way to football, eight or ten of us gathered on Sunday afternoons in David's yard to play tackle football, with the sidewalk as the fifty-yard line. I made sure of tackling David as often as I could, driving his knees into the morass, slamming him to the ground until my coat was covered with dirt and the shredded remnants of dead damp leaves and streaked with the green blood of the mashed and dying lawn.

CHAPTER 19

Not to be born is best.
SOPHOCLES

Late in October of 1960, after a short labor, Jo gave birth to a son, Sidney Ralph. He was a big strong infant, and he looked exactly like Carl. Except for obligatory visits to Hilt at Thanksgiving and Christmas, we saw little of Jo or her family until the following spring. Then, as Mother drove Elizabeth and me back and forth to Medford every month for orthodontic appointments, we stopped to visit Jo in Ashland. Carl still worked nights, so we didn't expect to see him on those trips, but one day in July, as Jo made instant coffee for Mother and Kool-Aid for us, Mother looked around the apartment. A good deal of the furniture was gone.

"You might as well know," Jo said, following Mother's gaze. Carl had left her: He was staying with his parents. They were getting a divorce. Mother opened Jo's refrigerator and looked inside. "How long has it been since you had any meat?" she asked, and Jo shrugged.

Two hours later, we left Ashland with Jo and Sidney and Elizabeth in the front seat of the Ford, and me in the back seat with boxes full of clothes and dishes and Jo's suitcases, and Sid's stroller, crib, and high chair folded up in the trunk. Grandfather shouted at Mother for interfering, but she was unmoved. "I couldn't just leave her there with no money and no food," she said. "She told me she and Carl were trying to work things out," Grandfather muttered, glancing over at his grandson in the playpen.

"Well, I don't think that's going to happen," Mother said. "We'll have to think of something else."

In my twelfth summer, in 1961, I learned to make drinkable coffee. On weekends, I carefully filled the stainless steel pot up to a line just below the spout, then dropped the stem into the bottom of the pot and fitted the basket onto it before spooning in the coffee. I placed the flat perforated basket-cover and the lid with its glass center over everything and put the percolator onto the stove burner. As the water boiled beneath the coffee and shot up the stem and bounced off the inside of the glass top, gradually turning darker and darker, I learned to tell by color and smell when the coffee was done.

I made the coffee one Saturday morning, on a day when we were all going fishing. The picnic cooler sat on the kitchen table, slowly filling with potato salad and fried chicken and sandwiches and cookies and apples, while I waited for the coffee to perk. I filled the two stainless steel thermos bottles with hot tap water to warm them, then lined them up, now empty, on the kitchen table, just as the coffee was ready to pour. I pulled the coffeepot off the back burner and turned around, colliding with Mother, who was swishing between sink and refrigerator, table and stove, behind me. I had made a full pot, and as I turned around and bumped her, the handle slipped in my hands, tipping the spout forward and sending about a pint of boiling coffee down the side of Mother's apricot-colored housedress.

The dress fit her torso tightly, and as I leapt backward to avoid the boiling liquid cascading onto the floor, Mother began screaming. She whirled in a tight circle, frantically clawing at the side zipper, trying to pull it down. Scalding coffee clung to her skin, held there by the dress and the nylon slip beneath and the hot metal zipper. By the time I had dropped the coffeepot into the sink and turned back to her, Daddy had rushed in and was swiftly ripping the zipper down and peeling the dress off of her shoulders. But she still kept screaming. Daddy flung me a glance as I cowered next to the sink. "It was the coffee," I got out, and I saw his face redden and his eyes bulge and glitter, while his lips drew back from his long teeth.

"I've *told* you to be careful with hot water on the stove!" he yelled, while inside my head a voice screamed silently for everything to please stop, for time to run backward, for this not to have happened yet, please, please, let it not have happened. For of course Daddy had warned us about hot water. But he was thinking of Tommy, not of Mother. From the moment Tommy came home from the hospital, Daddy brought Elizabeth and I to the stove and deliberately turned all the saucepan handles toward the back. When Tommy started crawling and then pulling himself up on the furniture, he must not be able to grab a pot handle and pull hot liquids down on his head. And at the thought of that happening to our little brother, and at the thought of what Daddy would do to us if that happened, we had shuddered, and had never, from that day to this, allowed a pot handle to point the wrong way.

Even as Daddy began stripping off her dress and her slip and unfastening her bra, Mother was flinging herself through the door of the master bedroom and onto the double bed, as though seeking refuge. While Daddy flung a sheet over her and ran for the bathroom to get wet washcloths to lay over her seared ribs, Elizabeth, who had been standing across the kitchen smearing a piece of cold toast with peanut butter, slipped silently out the back door. I wanted to follow her and could not. I wanted to go out that door and grab my bike and fling the back gate open and knock the bike's kickstand up with the toe of my sneaker and be mounted and racing down the alley and onto the blacktop and coasting, coasting, for the center of town and the cool shade of the cottonwoods in front of the store. But I stood there in the kitchen, frozen with horror and the knowledge that this was what I got for thinking I was so great at cooking. I had been so proud that morning, and I had felt so superior to Elizabeth, who was sulking because she wanted to go up to Adobe Street and play with Lynn that morning. Now, how I envied her, out the back gate and down the alley, scot-free and guilt-free, and the furthest person from Daddy's mind at the moment.

For an endless half-hour, Daddy shot curt orders at me through the open bedroom door. Pound up some ice cubes in a dishtowel and bring them here. Get some aspirin. Bring a glass of water. Call your grandmother. Daddy sat at Mother's side on the bed, holding cold wet compresses to her white skin, gently removing her glasses and setting them on top of the bookcase-headboard, slipping an extra pillow under her head. Mother faced the wall and never looked at me, never spoke to me. She lay on her unburned left side, her knees tucked up, her head bent back, absorbed in the pain. She moaned and shivered and gripped Daddy's arm hard with one hand. I came out of the bathroom with the aspirin bottle and saw her reflection in the long mirror above the low bureau, her face white, her eyes squeezed shut. I handed Daddy the bottle and walked out to the kitchen and got the glass of water and picked up the towel full of ice shards in the sink and carried both of them back to Daddy, handing them to him without a word. As I picked up the wall phone in the kitchen and began to dial the familiar number, I heard Daddy murmuring to Mother, "Come on, Barbie, take a couple of aspirin."

I had dialed three digits when I heard the clack of the back gate opening and ran to the back door. Grandmother gave me a tight little smile and trotted past me into the bedroom. I padded behind her, then continued into the front room. I heard Mother rousing herself, her voice rising into a dismayed, apologetic sobbing. I heard the words "second-degree burns" as I paced around the front room, then into the bedroom I shared with Elizabeth, then through the bathroom, until I stood in the doorway of Mother and Daddy's bedroom, watching as Grandmother and Daddy bent over Mother, trying to relieve her agony. Daddy looked up at me. "Get out of here," he said, and I went out at last onto the front porch. Through it all, Tommy had slept on in his crib.

I sat on the top steps in the hot August sunlight, surrounded by the smell of old wood and the noises of bumblebees in the hollyhocks. It seemed a long way down to the ground, and I looked out at Bailey Hill, at eye level with me, dreaming in the hazy quiet of the morning. I sat on the steps and watched the ants and wished I were dead.

Fear was an old acquaintance to me now, but it could still surprise me. The panic attacks that had started when I was eight still occurred, in situations where I least expected them. Fear lay in wait, always ready to pounce. But the feeling that came to me now on the front porch went deeper than fear, beyond it into a black despair that sat on my shoulders, that I could not escape, or outrun, or outlive. And as I sat there I knew that the despair was deserved, entirely my own fault. I had done something so bad I could never atone for it. Mother and

Daddy—especially Daddy—would never forget this, never forgive me. In that hour, as I sat alone on the porch, I was afraid Mother would die, but I was more afraid of the thought that I must live.

I sat on the steps and saw only one way out. I knew where the .22 rifle was, the only one that I had ever fired. I knew where the other rifles were, too, and the revolver, but the .22 would be enough. If I rested the shortened stock on the ground, I thought, I could just reach the trigger with my finger while aiming the barrel at my forehead. I knew how to work the bolt action, insert a cartridge, and slam the bolt home. I could do it. But all the guns were in Mother and Daddy's room, and much as Daddy might wish me dead right now, I knew he wouldn't let me get the .22 out of the closet under his very nose.

I stood up and leaned over the porch balcony, but as high as it looked from here, I knew it wasn't high enough. I had jumped off of it many times and never gotten so much as a sprained ankle. Poison? But I wanted oblivion, quick oblivion, not suffering, and I had a hard time swallowing pills anyway. Out of ideas for the moment, I sat back down, and began to cry, and before long I knew that what I really wanted was not to die, but to become forever invisible.

I heard the front door opening and braced myself for Daddy's wrath, but instead Aunt Jo came out, holding Tommy, gently patting his back. Tommy lifted his head from her shoulder and looked around at me, his face creased with sleep, his T-shirt and terrycloth jumper stained, as usual, with urbled milk. He was two years old and drinking from a cup most of the time, but he still threw up about half of what he ate. He grinned and flapped his hand at me, and I held it and then took him onto my lap as Aunt Jo sat down beside me.

Elizabeth reappeared, cruising the sidewalk below us. She leaned her bike against the fence, gingerly opened the gate, and came up the walk. She sat below us on the long flight of steps for a while, then slipped back out the gate and coasted silently away. I wanted to be her, to be Elizabeth, who was never afraid, who could stand up to the worst and not give in, and most of all, who had not done this thing to Mother.

By the time Mother's burn was bandaged and the pain somewhat abated by some old codeine pills from Grandmother's medicine cabinet, and Grandmother and Aunt Jo had gone home, I realized that

Mother was not going to die, after all. Whatever I was, I was not—at least technically—a murderer, and if I had not been exactly forgotten, I was, at least for the time being, not missed very much. My bike was near the back door, too close to Mother's sickbed for comfort. In the absence of anyone requesting my presence, I hopped over the front fence and ran on my toes, quietly, down to the church, then swiftly across the county road and behind the garage building just below Bob Trinca's service station. I worked my way up the dry bed of Hilt Creek and sat crouched in the galvanized culvert under the school's driveway for a few minutes, listening for the sound of my name being shouted. After a while I crawled out and ran for the school playground. I sat on a swing for a while, then walked slowly down the back alleys to Grandmother's house.

Grandmother looked sadly at me, but didn't rebuke me. Aunt Jo was playing "Isle of Capri" softly on the old piano, and I wondered that I had never before realized what a melancholy tune it was. Grandfather, crouched in front of the radio in the living room, was not disposed to be charitable about the day so far. "That was a pretty stupid stunt you pulled," he boomed at me through a pall of cigarette smoke, over the voice of the baseball announcer.

"I didn't mean to . . ." I began, but he had already shifted his attention back to the Giants game. Grandmother fed me a bowl of bananas and milk, gave me a gentle shove out the back door, and locked the screen door behind me. She looked frightened and worried, and I felt myself grow hot with fear, this time for Mother. Was she going to die, after all? In books, I had noticed, people sometimes survived one crisis only to Go Into a Decline and Die, like Beth in *Little Women.* Perhaps Grandmother knew that Mother was going to die, and that was why she was locking me out of her house.

"Run home," she told me. "Quick, like a bunny."

I ran for the front yard and clattered down the sidewalk. Next door, on the top step of what had once been Ohlund's house, Lester Chase sat in his undershirt, a beer bottle dangling from one hand between his knees, the other hand buried in his black curls, clutching his head.

"You're supposed to be at a party at Barron's this evening," Daddy said to me without preamble as I came through the kitchen door. His arms were deep in soapy water in the sink. Mother was asleep at last. I looked up at him in astonishment. I would not even have dared to mention my invitation to Mimi's party after what had happened this morning. A party, I felt, was the last thing that would be allowed me for a long time. I had pictured myself going to Mimi in a day or two and giving her the birthday card and the present I had picked out for her and wrapped a week ago, with some lame excuse.

"If you'd been around earlier, I would have spanked you already, but now you'll have to go to the party first and think about what you did. Mr. Barron asked me yesterday if you were going to be there, and I told him you were. So get dressed and go, and I'll deal with you when you get home."

I changed into a dress and clean socks in my room, then scurried out the front door so I wouldn't have to walk through the kitchen again.

Mel Barron had been Hilt's resident manager since 1956. He was much younger than old Mr. Baumann; he wore golf shirts and built birdhouses in a wood shop out behind the big Company house. He went to school plays and his wife led Hilt's Girl Scout troop. Their daughter Mimi was in the eighth grade, slender and sophisticated, already wearing lipstick. She had a horse, a beagle, a huge doll collection, her own record player, and a vast collection of 45 RPMs, which she let us play as we sat cross-legged on her frilly bedspread, under the glassy stare of row upon row of plastic faces.

Mimi's house had a wide veranda, shaded by old horse chestnut trees with inexhaustible supplies of hard, smooth, lethal missiles. To me, the house was a palace, its yard an estate, and I had been looking forward to her party all week. We would play records and dance on the veranda like high school kids, not play stupid children's games like pin the tail on the donkey.

As I neared her house, the summer evening was fragrant with the late roses on the fences. I saw the lawn chairs and the covered tables bearing cookies and potato chips and a huge crystal punchbowl full of pink lemonade. There were even boys there: Pat Trinca and David Simmen, Billy Wilson and Danny Burns, David Powers reading a hot-rod magazine in a chaise longue, and Ronnie Chase, who looked like Ricky Nelson. Janet Cavin and Carleen DeClerck kept a proprietary eye on Gary and Larry, their boyfriends from Hornbrook.

Periodically, the boys whipped combs out of their back pockets and gouged them through the long greasy locks plastered to the sides of their heads.

The record player was turned up loud, and beside it—oh, blessings!—was every single record Mimi owned, so that I could stand by the record player and wallow in reading the album covers, to cover my embarrassment at being both late and in the presence of so much sophistication.

We ate birthday cake and little crackers with Cheez Whiz squirted on them, and Mr. Barron dipped punch out of the crystal bowl and joked with us. After a while some of the older kids began a desultory shuffle over the wooden veranda floor, an activity that increased as the darkness came. I stood up with my classmates Phyllis Jordan and Helen Shepherd and we danced all together, and I smiled and laughed and tried not to think about the awfulness of the day and what waited for me at home that night.

By nine o'clock the undigested Cheez Whiz was warring with the sickly sweet punch and the rather heavy cake in my stomach, and I felt sick. I lay back, by myself, in the dusk, on one of the chaise longues. I won't throw up, I said to myself. I won't. I can't. I can't go running into a strange bathroom and throw up. My heart began hammering, and my stomach felt worse.

I couldn't tell anybody how I felt. I had defended Daddy so often to the other kids, who knew how different my life was from theirs, in the full knowledge that a parent's faults, if admitted, were always ascribed to the kid. For better or worse, kids in Hilt identified with their parents. A bad parent was a reflection on you. So kids always stuck up for their parents in front of other kids. Oh, a girl could complain a little about her mother—that showed a proper degree of maturity—but complain about your father? Never.

I couldn't even talk to Grandmother about this one—she might tell me that she, too, thought I was a terrible person. Grandfather already seemed to think so.

Perhaps I could confide some of it to Aunt Jo. She always listened to me as though I were another adult, a real person, although lately even she seemed to be distracted by something, not really paying attention when I talked to her.

As the sky slid into full darkness, and Janet and Carleen danced closer and closer to Gary and Larry—or was it Larry and Gary?—I tried to take deep breaths, which made my stomach feel better, and lie back on the chaise longue, which also made it feel better. Somebody had put a Johnny Horton album on. He was my favorite singer, and he was dead in a car crash, and his voice was singing about springtime in Alaska, whisperin' pines, and a darlin' who was gone. The tears prickled behind my eyes in the dark and poured down my face in safety, while I let them come and felt very, very sorry for myself.

By the time the party broke up at ten o'clock, I was pretty well dehydrated. I scrubbed the tear marks from my face, located Mimi and her parents, and said my well-trained thank-you's. I walked out the front gate and down the sidewalk and past Grandmother's house, where the lights were all out except the one in Jo's bedroom. The cottonwood tree groaned and thrashed in the warm night wind. The long branches managed to look menacing under the streetlight, pale and shivering, the shadows creating strange figures that refused to come into focus. I thought I saw an upright figure far behind the tree, next to a back window, stationary behind the gyrating shadows. But under the dim streetlight and the rising moon, I suddenly didn't want to linger there. I began to run, my footsteps thudding up the sidewalk, despite what waited for me at the top of the hill.

I went the long way home, trotting past Alphonse's house, rather than crossing directly over to the store and going straight up the steep sidewalk beside it. I followed the county road up to the church, then turned and walked up the alley. Behind Johnson's house, their old orange-and-white spaniel growled at me from behind the fence. I ran in our gate and walked up the two steps to the back porch. The porch light was on, and I pulled open the screen door and pushed through into the laundry room, where I stopped and listened before opening the door into the kitchen.

The light was out in the kitchen, but the living mom lights were on, and Mother was sitting up in her nightgown in the maple rocking chair, a cup of tea on the end table beside her. Elizabeth was already in bed, and Tommy was asleep in his room. As I came in, Mother stood up without looking at me, and silently and carefully closed the door to the nursery. Daddy was sitting in his easy chair beside the front door, reading *Sports Afield.* On the cover, a pheasant flew up in front of a hunter with a leveled shotgun. Daddy rose and beckoned me after him. I followed him through the kitchen and into the back bedroom.

Daddy picked up the stick we called The Paddle from its resting-place on top of the dresser and sat down on the edge of the bed, on top of the tan chenille spread. "Okay," he said.

I knew the routine. I approached him, lifted the skirt of my dress, and lay down across his knees, my heart slamming into my throat. His arm came down over the middle of my back, holding me down, and I reached out with one hand and clutched the hem of the bedspread, staring down at the pattern on the linoleum and at a dust bunny just under the bed that Mother had missed in her last vacuuming. I told myself, as I had done many times before, that *this* time I would be brave, *this* time I would not yell and scream for him to stop, *this* time I would be stoic and think of something else: the times tables, Roger Maris's quest for the home-run record, state capitols. But there was a protocol to Daddy's spankings. If we managed to delay our cries somewhat, Daddy thought he wasn't hitting us hard enough, and he would strike harder and harder, for as long as it took to make us yell. If we yelled too soon, he would shout at us to be quiet, that he wasn't hitting us that hard, and would pause momentarily before delivering a vicious series of blows, which left us sore for days. And if we sobbed too long after a whipping, he pounded on our bedroom door and told us that if we didn't shut up, he'd come in and give us something to really cry about, as we gulped and hiccuped and buried our faces in our pillows and struggled to control our breathing.

I made it to the fifth blow, this time, having had so long to think about it and form a resolve to be tough. But the sixth blow and the seventh were too much, and I began to yell, as the blows came harder and harder. This, as I had feared, was one of the Really Bad Ones. By the ninth blow, I was screaming, and I flung one hand back to try to shield my buttocks, which might as well have been naked for all the protection my panties afforded.

The stick Daddy wielded, a sturdy piece of fir about fourteen inches long and two inches wide, had been whittled flat, with a handhold near the base, and carefully pared down with the small blade of Daddy's pocketknife, to remove splinters. But the paring had also sharpened the edges of the stick, and I yelped as my little finger got in the way of its descent, exploding in a pain like fire before going numb. I snatched that hand back and threw the other one out behind me, and it got hit, too. Without an arm to brace myself against the floor, my head and shoulders began to slide down. Daddy grabbed my hands and held them in his left hand, hoisting them above the small of my back. The blows resumed. I couldn't breathe, or think, only emit throat-tearing screeches. He was going to kill me for sure, this time. In a last, desperate effort to make him stop, I interspersed the shrieks with words—"DADDY—DADDY—DON'T—DON'T—DON'T—IT'S—JUST—THAT—I—LOVE—YOU—SO"—the words came out in a disjointed croak, but as soon as I had them all out, the blows came on harder than ever, ten or twelve of them, delivered with more force, but less aim, so that I knew he was finally tiring, and the last ones fell willy-nilly on my bare thighs, as I simply lost track of time, reduced as I was to frantic writhing and screaming. When he let go of my wrists at last and stood up, his knees pushing me away from him, I dropped in a heap on the cool floor. Nearly stunned, I crawled as fast as I could through the bathroom door and reached up to pull it shut behind me.

I sat on the bathroom floor, my weight on the side of one hip, cradling my numbed fingers in my armpits, and sobbing with the all-encompassing sobs that seize the diaphragm and take on a life of their own. I heard my parents' voices in the front room, low pitched and high pitched, talking. I hooked an arm over the cool, smooth enamel of the bathtub, and turned the cold water on full force into the tub, and kept crying. I tried to choke the sobs back, and gulped, and choked, and coughed, and for a moment thought again that I would throw up. I won't throw up now, I told myself, as sweat broke out on my forehead and the room grew hot and cold by turns. I pulled a wad of toilet paper off the roll and held it under the tap, squeezed it out a little, and applied it gingerly to the erupting welts on my thighs. The cold water stung. I put my still-numb fingers under the tub tap, and they slowly began to feel better. I splashed water on my face and looked up, out the window, at the dark. I stood up slowly, saw my toothbrush and picked it up and squeezed toothpaste onto it and began, mindlessly, to brush my teeth.

I heard Mother and Daddy turning out the front-room lights and moving through the kitchen and into their bedroom. I jumped for the bathroom light, noiselessly switched it off, and slipped out through the other door into the darkness of our bedroom. Shaking, I sat on the edge of my bed and undressed, pulled my nightgown over my head, and slid under the covers. I heard Elizabeth turn over in her bed across the room.

"Are you okay?" she whispered tensely.

"Yeah," I answered. I heard my parents taking their turns in the bathroom, and I lay in the dark, afraid. But no one came in, or disturbed us. Wide awake, I began to tell myself an episode in a story I had been spinning in my head at night for about a year, a story about a family of chipmunks who lived in a clean dry hole in the ground, beneath a juniper tree. I was just dozing off when I felt Elizabeth crawl into bed beside me. I scrunched up closer to the wall to make room for her, and she threw an arm over my waist, and we slept, curled up together like spoons for the rest of the night.

CHAPTER 20

And human love needs human meriting:
How hast thou merited—
Of all man's clotted clay the dingiest clot?
Alack, thou knowest not
How little worthy of any love thou art.
FRANCIS THOMPSON

Daddy had to work on the fire desk the next day. He was gone by the time we were awake. For once, Mother hadn't fixed his breakfast or packed his lunch. When I finally crept out of bed, Mother was in the kitchen, wearing one of her old maternity blouses and no bra, but looking much better. She placed a bowl and a box of cereal on the table in front of me, and I felt the deadly silence that meant I was still in trouble. I ate the Cheerios, then made myself a glass of Tang and a piece of peanut-butter toast and escaped to the lawn to eat them.

I sat on the damp grass, which felt nice and cool through the thin corduroy of my pedal pushers. I had put them on after first taking a cautious look at my buttocks in the long mirror on the closet door. They were a mass of purple, the color really startling, continuing down onto my thighs. "Wow," said Elizabeth, awestruck, as she looked at the violet-black bruises. I knew I wouldn't wear shorts today; I didn't want Mother to see the marks.

Across the fence, I saw the top of Ruthie's red head moving along the side of her house, her head bent as she watered her dahlias with a hose.

The Johnsons must have heard the noises I had made. Their windows, and ours, were open to the soft August night. But they would say nothing, just as no one ever said anything when Mr. and Mrs. Benson across the street had their frequent late-night screaming matches.

I lay back and watched the morning sunlight filtering through the dancing, almost transparent, leaflets of the locust tree. I wanted to climb into that other world and stay there. But this tree wanted no one to climb it. We had tried many times, my sister and I, but the vertical limbs and the thorns always drove us back before we had reached the second tier of branches. There was no place to sit there, no place to rest. The beautiful world of the yellow-green leaflets, where small birds whisked in and out of sight, remained firmly out of our reach.

All that week, Daddy left early and came home late, as he tried to stay one jump ahead of the loggers. In the back of my mind, I knew that something remained unsettled between us, something that had nothing to do with Mother's burns. *Those* had been paid for, with my bruises. No, it was what I had *said* in my desperation that hung in the air. Sometime soon, I knew, I would get a lecture, probably about how bad it was to try to avoid a punishment that I richly deserved.

I dreaded the weekends now, for sooner or later Daddy would be home for an entire day with us. When he had to work a second Saturday in a row on fire patrol, I seized the opportunity to ask Mother if we could stay overnight at Grandmother's and, rather absently, she said yes.

Perhaps I wanted Grandmother to see my bruises. I certainly didn't object when Elizabeth told her about them. She clucked and looked at the marks, now fading to yellow around the edges, and dabbed witch hazel on them, although the welts were almost gone. While Elizabeth snored on the other side of the old double bed in the guest room, drooling onto her stuffed panda, I lay awake clutching my cinnamon teddy bear, listening as Grandmother talked to Jo in the bathroom. Jo was bathing her ten-month-old son in the tub, and if I closed my eyes, the sound of Jo's voice let me pretend that I still lived here, and that everything was as it had been. But it was not.

As Elizabeth and I walked home that Sunday after breakfast at Grandmother's, I wondered why I couldn't be more like Tom Sawyer, or Penrod, who took their lickings and then forgot them. Why couldn't I be brave like them? Over the years, I had tried to escape the pain of Daddy's whippings. Once, I stuffed wads of toilet paper down my underpants and tried to fake the yells. But of course Daddy felt the padding and sent me into the bathroom to remove it, and in the end the whipping was not so very bad. His blows lacked their usual fervor, and, sneaking a look back at him afterwards, I saw him biting his lower lip, his eyelids crinkling. He was laughing.

But now it was not a whipping I feared. I had hoped for a forest fire yesterday, or something that would take Daddy into the woods again, but my luck had run out. Daddy met us as we came in, and I saw the kitchen chair sitting in the middle of the front-room floor, waiting for me, and knew that I was in for it.

I hoped for a lecture about safety, and thinking, and being careful, and braced myself to look contrite and repentant. This was not hard, for I was both, and now so terrified of scalding someone that I carefully checked behind me before moving anything from the stove to the sink. That was what I hoped, even as I knew that something else was brewing. And I was right, although when Daddy began to speak, I was utterly surprised.

My sin, it turned out, had nothing to do with trying to escape punishment. Instead, it was much blacker, and as he described it, the bottom fell out of the little platform I had painfully constructed for myself above the abyss.

I had told him that I loved him, he began, in order to get out of a whipping. That was bad enough, because I deserved it, and should be willing to take it. But what that plea revealed about me was worse—I was a liar. I didn't love him. I didn't love anyone. If I did, I would obey him and do my chores without being told and help Mother without being asked. I didn't know what love was. I was a selfish child who had done nothing to deserve love herself. People who loved other people showed it by working for them and being considerate. Then they would deserve love themselves. They didn't *tell* people they loved them. Telling people you loved them, or kissing and hugging them—that wasn't love.

My mind froze. I kept my face rigid, a mask of attention that carefully revealed no emotion. Oh, Tongue, I said in my mind, you're so lucky, hidden in my mouth. He can't hurt you. You're safe in the dark, behind my teeth. I wish I were you. You can waggle and turn and he can't see you, can't beat you. You can hide. Be happy you're safe, because I'm not.

There's no such thing as falling in love, Daddy's voice continued, as though from a great distance. Real love has to be earned and learned. I looked from Daddy to Mother, sitting primly in her rocking chair nearby, her ankles crossed, her hands in her lap, her back straight. Her face held no sympathy, no softness.

I was unnatural, Daddy said. Someone who claimed to love her parents, but didn't obey them, wasn't normal, or even sane, and such children, if they didn't shape up, were taken away and locked up in a mental hospital.

My heart began to slam against my hot stomach. I could no longer feel my feet. My head seemed to be floating above my body, light and dizzy. Three years ago I had been terrified that my parents would desert me. I had gradually convinced myself that this wouldn't happen, but now I heard Daddy saying that I was crazy, and that crazy people were sent away. It was one thing to be eight years old and afraid of being abandoned at a gas station; it was another to be twelve, and wearing a training bra, and hear Daddy calmly threatening to send me away to a mental institution because I didn't love my parents.

I tried to make sense of all this. I had thought I loved my family—no, that wasn't true, I never thought about it at all, for who thought about something so obvious? But I had *said* so, at exactly the wrong time, and now Daddy was telling me that for me to have said those words was wrong. Not only did I not love them, but also I literally *could* not, because I was selfish and lazy and inconsiderate, and such people did not love.

The foundations of the earth rocked. If I loved no one, if my actions gave the lie to any assertion that I did love, then no one could possibly love me back.

"You don't know anything about love, and you certainly don't love your Mother or me," his voice was saying now, still far away, but growing louder. His pale and angry eyes looked bleached in the afternoon sun that poured through the twelve-paned windows. "If you did, you'd obey us and do what we want you to do, and I wouldn't have to spank you. So when you say you love us, you're not telling the truth. And if you don't love anyone, why should anyone love you?"

The words left me no escape, no way out. I couldn't even say I was sorry. I'm sorry I don't love you? Words spoken in pain and panic had changed everything and left me exposed for what I was—selfish, self-centered, no good, crazy. A good person would want to help her parents, want to obey them. But I wasn't a good person. I wanted to ride my bike and read books and play basketball and fly kites off of Watertank Hill. I wanted to be free and not do housework. I didn't want to grow up. What Mother said sometimes when she was angry with me was true, then. I *was* like Dick Johnson, after all.

Two months before, on a weekend in June, Dick Johnson had briefly re-entered our lives. He came to sign, at long last, the adoption papers Mother had been urging on him for four years. He had always refused and kept on sending the support checks, even though Mother no longer needed them. But just after my twelfth birthday, he arrived, with his wife, Pauline, and their three-year-old son, Tommy.

As the two Tommys played on the floor, one thin and blond, the other plump, with black curls, the adults made uneasy conversation. Pauline was polite but distant, Dick shy and uneasy. A strained luncheon ensued. After the meal, Dick asked to speak to Elizabeth and me alone, and we went out into the backyard with him.

Mother had primed us for this moment. When she had told us that Dick was coming to sign the adoption papers, she seemed nervous about what he might ask us and about what we might say to him. "If he asks you if Daddy ever spanks you," she told us, "You should say, 'Well, sometimes when we're *very* naughty.'"

We nodded. I was even a little miffed that Mother had thought it necessary to warn us about possible questions. I knew how important this adoption was to Mother. I wasn't about to hurt her feelings by trying to mess it up for her. Elizabeth and I could keep secrets—any secrets, and from anyone. We could keep our secrets, and Mother's, and Aunt Jo's. We wouldn't squeal.

Dick got down on one knee before us, on the lawn beside the vegetable garden. I don't remember everything he said, except that he told us he had always cared about us, and that he was letting John adopt us because he thought it would be the best thing for us. We nodded. He said that he wanted to keep in touch with us, and that if we ever needed any help from him, we had only to ask. We nodded again, and he hugged us both, and then he walked abruptly away from us, out into the alley, and strode up and down behind the cars for a few minutes. Then he came back, and we walked back into the house together.

When they were ready to leave, I ran to get my little camera and took a picture of the three of them standing by the front gate. It didn't turn out very well, so that in later years, when it was still the only picture I had of him, I could tell only that he had been slender, and blond, with long thin features. He remained a stranger, a stranger hated by our mother. That he should not want us seemed logical to me; that Daddy should want us was, as always, illogical. But we knew that Mother would regard any future contact with Dick as disloyalty on our part, would see it as a rejection of the efforts she had made on our behalf, to bring us a better life by marrying John Brannon. For that reason alone, we knew that we would never write to Dick again. We would try to forget him, and sometimes we would almost succeed.

So I hadn't told Dick about the spankings, and the rules, and the punishments. But I could have, I thought, as I looked over at Mother—Mother, who now would not look back at me. I sensed, at that moment, that she was weighing me against her other two children, weighing me against the loyalty she owed to her husband for marrying her and supporting us all. Against that heavy debt, one careless and disobedient girl could not begin to tip the balance.

Daddy was still talking. My ears popped the way they did coming down from the Siskiyou Summit. "Showing people physical affection," he said, "that doesn't mean anything. If you love somebody, you show it by working for them. Do you understand?"

I nodded. But Tongue yelled, silent in the dark, "*You* hug and kiss Mother when you come home and every night after supper. But maybe that's different, something adults do and tell us not to do."

At the end of the afternoon, I found myself in my room with the doors closed. After a long time, Elizabeth came in, and then it grew dark, and we slept. I woke a few hours later, straight out of a dream in which I was wetting the bed. I had not done *that* for a long time, but I lay rigid in fear for a few minutes, until I was sure it had been only a dream. I got up, silently unhooked the window-screen latch, and slid out onto the porch. I padded across the porch and down the steps, squatted on the lawn, and peed.

I looked up at the summer stars and wished that I could go there, to another planet, another galaxy, where the afternoon's lecture would stop playing itself, over and over, inside my head. I wished that I could erase the pictures in my mind: of Mother looking out the window, of Elizabeth on the far end of the couch, chewing on an already soggy white handkerchief. Elizabeth, who I might have expected to show a sisterly glee at the trouble I was in, had instead looked worried, as though what was happening went far beyond her understanding. You and me both, kid, said a voice in my mind, with Jo's inflection.

"And one more thing," the endlessly looping recording of Daddy's voice repeated in my mind as I crawled back through the window, "You're not to run crying to your grandmother and upsetting her about this. She has enough on her mind right now. She doesn't need you going down there and making more work for her. I don't want either of you to go down there for two weeks."

At that part, Mother's eyes had at last flickered upward. They must not have discussed that. She compressed her lips and looked away, away from me, away from Elizabeth, away from Daddy. She looked at Tommy, playing unawares in the middle of the floor.

Did Mother and Daddy love Tommy? How could they? Tommy was a baby. He drank his milk and threw more baby food around the room than he ate. He dirtied his diapers and screamed all night and stuck everything he picked up directly into his mouth. He was two years old. He didn't make his own bed or help with the dishes, and he had never done a considerate thing in his whole short life.

And if love was something that grew from doing things for other people, then that meant that Daddy must not have loved Mother when he married her, for what had she done for him before they were married? She didn't start scrubbing and cooking until later. Did he come to love her after he married her? That was what I heard him saying. Did Larry and Gary hold hands with Carleen and Janet because they were hard workers with good manners? I doubted it. Boys didn't think about stuff like that. Nobody did, except Daddy.

Did I love Tommy? I got up at three in the morning when I heard him start fussing and walked him back to sleep, rubbing his back and singing the old nursery rhymes that Mother once sang to us. I changed his diapers and spooned strained vegetables toward his face. Did I love him? After today, I no longer knew. I only knew that I never wanted to hear the subject mentioned again.

Out by the old ballpark, a coyote yipped, then howled. An owl hooted in a white poplar tree across the street. The night breezes blew through the open windows, warm and then cool, smelling of juniper and sage and wet lawns and cut grass. I wondered what would happen if Elizabeth and I woke up in the morning to find ourselves alone in Hilt, the only people left here, and everyone else gone. We could ride our bikes wherever we wanted, and come home whenever we liked, and not be afraid, ever again.

". . . and then he said not to come down here for two weeks!"

Aunt Jo patted my arm and looked away, up and out the kitchen window above the sink. Didn't *anyone* want to look at me anymore? I hadn't told her much about Daddy's lecture. I was afraid that if she knew too much about it, she might agree with him.

"Don't tell Mother about this," I hiccuped. "She'll tell Daddy." "No, I won't," Jo reassured me. "But you come down here anytime you want, and if John doesn't like it, too bad."

As it turned out, even as I sat telling my troubles to Aunt Jo at the kitchen table, Grandmother was marching up to talk to Mother. She had listened to us from her sewing room for a few minutes, then abruptly stood up and announced that she was going to the post office. We nodded at her as she went out the front door, slamming it behind her.

"Is she mad at me, too?" I asked, and Jo shook her head, no, and got up to get me more lemonade and another molasses cookie.

When Mother opened the back door to Grandmother's knock a few minutes later, she was surprised to see a look of fury on her mother's face. "What's the matter?" Mother said. "Come inside, Mama."

"No," Grandmother retorted. "I'm not going to come in. I just want to tell you one thing: if you go on like this, you are going to lose those girls. Don't say I didn't warn you!" And she turned on her heel and marched back through the gate, banging it behind her, not turning to re-close it when the latch didn't catch. Mother ran down the steps and stood amazed, staring at her mother's rigid back retreating down the alley.

I was at Grandmother's a lot over the next few weeks, going on foot so my bike wouldn't give my location away. In the drowsy afternoons, while Grandmother napped in her room and Jo dozed on the sofa, I opened the bookcase. Zane Grey would never tell me I was selfish; Nancy Drew's father was generous and approving and never hit his brilliant detective daughter. I took books out to the old white poplar and climbed over the rough gray limbs into the smooth whiteness of its upper branches and sat hidden by the masses of quivering leaves, with a book in my lap. For years now, reading had been my drug, one of the few anodynes that dulled my self-hatred. Sitting sheltered here, out of sight, I could forget that I wasn't normal. The old tree at least liked me, and the people and animals in the books liked me, and they all remained with me, friends and confidants that no one else could see, long after I climbed down and walked up the hill.

Mother's burns healed slowly. Sometimes I saw the scar as she dressed or undressed. By the time I could hug her again without hurting her, I was ashamed to do so, and I did it less and less, especially where Daddy could see me.

Elizabeth and I came to a new understanding of our relationship. We never talked about it, but we started looking out for each other. We never tattled on each other again. We backed up each other's stories, covered for each other. We never again forgot—in spite of fights and hurtful words—that we were sisters.

I never tried to hug Daddy again, or to sit on his lap, or to kiss his cheek. I never again mentioned the word *love* in his presence. Years later, when I went away to college, we shook hands.

Almost forty years later, I know what he feared. He feared that we might, without a great deal of discipline and hard work, succumb to the same temptations as Jo, and with much the same results. With me, he was to succeed only too well; with Elizabeth, he succeeded not at all, for she could never have become like anybody except herself.

CHAPTER 21

> He that studies books alone will know how things ought to be; and he who studies men will know how they are.
> CHARLES CALEB COLTON

We had done almost everything together once, my sister and I. We took ballet lessons together, and at recitals we always performed together, doing numbers like "I Take My Doll to Dancing School," in which we carried identical dolls, dressed in miniature ballet shoes and tutus. We went trick-or-treating together and were immediately recognizable by our identical costumes, one ghost simply shorter than the other. We learned the same songs, recited the same rhymes, played with the same kids. But after Mother married Daddy, we dealt with our changed situation in totally different ways.

The years with Daddy had taught me my limitations. For one thing, I was not nearly as tough as I had once believed. In Daddy, I had for the first time in my life met somebody who chose to bring his advantages—bigger, smarter, older—to bear upon me, for the purpose of making me behave in a certain way. I reacted to the loss of control over my life—to the new discipline and rules and punishments—with panic attacks and nervous tics and by withdrawing from human beings, who had betrayed me. I drew closer to trees and meadows, sunsets and clouds, lizards and frogs. I fled to books, which never yelled at me, or reminded me that I wasn't pretty.

But Elizabeth, my skinny, delicate, blonde sister, found another way. She forged new ties with other families and their children. It's just one man's opinion, her instincts told her, as Daddy spanked her. She faced firmly outward, no more able to turn away from people than to stop breathing.

When Daddy ordered Elizabeth not to go crying to Grandmother after a spanking, she shrugged and pedaled up the hill to Lynn Marin's house on Adobe Street. Lynn was, from the first grade, her best friend. Elizabeth often stayed for supper with the Marins. At first, Daddy thought that she was taking advantage of their hospitality, and he spoke to Tony about it, telling him not to let Elizabeth just invite herself any time she wanted. Tony assured him that Elizabeth was welcome anytime, that she was a joy to have around. Daddy stared at Tony, then told him to just chase her on home if she made a nuisance of herself.

Elizabeth's undeniable charm had never worked on Daddy. She was, he believed, a manipulator of others for her own purposes, and it amazed him that nobody else noticed this, and that others loved her for the very qualities that infuriated him.

At ten years of age, Lynn and Elizabeth were sneaking cigarettes from Inez's purse. Out in the alley, they sat with their backs against the warm gray woodshed wall and lit up. Hacking and choking, dizzy, they passed the cigarette back and forth, until there came a day when the nicotine no longer bothered them, but instead made the whole world better. Two blondes, one plump, one slender, their long yellow hair over their shoulders, sitting and puffing in the long afternoons. As Elizabeth learned to blow the smoke out through her nostrils, and make smoke rings, she ceased to be Elizabeth and became Liz, the Invincible.

"Louise," she said to me one day, not long after the worst day of my life, to which she assured me she had been a sympathetic, if helpless, witness, "I don't want to be called 'Elizabeth' anymore." She spoke in a perfect imitation of Inez's husky drawl. "I want to be called 'Liz.'"

"Why?" I asked. She sat on a litter of wood chips in the late August sunshine, her back against a stack of fir rounds. She tapped the ashes from the end of a Winston, stolen, this time, from Grandfather. He was trying filter tips. I tapped tentatively with the axe on a pitchy chunk of fir, splitting it into kindling.

"Because Mom *hates* it," she said. And that was true. Mother had already heard other children, and even their mothers, call her that. Mother thought it sounded coarse, fast, sleazy, as though her small blonde daughter were some kin to the disreputable Liz Taylor. And that, of course, was precisely why Elizabeth loved it.

"Okay, *Liz,*" I agreed, willing to join the game. "Be sure not to leave that butt here."

Liz's talent for sheer, raw stubbornness was legendary in our family. The major stage for its exhibition was the supper table. Nobody at Grandmother's house had ever paid much attention to Liz's eating habits, or mine, for that matter. We ate as much or as little as we liked and left the table when we were good and ready.

By the time she was seven, Liz had grown a couple of inches and become thin, with a pale complexion that, against the backdrop of her long yellow hair, made her look green. She had little appetite. If she ate a cookie at three o'clock, she spoiled her supper, and under the new regime, we had to clean our plates, not just in order to get dessert, but in order to leave the table at all. But a Liz who wouldn't eat her supper was also a Liz who could wait everyone out. Eventually, I would choke down the last awful mouthful of liver, but Liz could sit at the table staring into space until ten o'clock. If pressed, she would choke down a couple of bites, then turn her head and toss everything back up, effortlessly, onto the floor. When Daddy spanked her for this, she threw up again, then had the dry heaves all night. He finally resorted to slapping her across the face and dragging her by one pale arm to her room, ordering her to stay in there for a week except for school and meals.

Liz was not a model prisoner. She didn't seem to mind her incarceration, for one thing. She rat-holed comic books under the mattresses and, if deprived of them, swung from the clothes bar in the closet, singing in a high reedy voice that drove Mother crazy. When a week in solitary had still not increased her appetite, Daddy was forced to cave—a little.

He still demanded that she clean her plate, but he allowed her to take tiny portions. When she failed to manage even those, he glared at her, forbade her desserts for a week, and told her to stay away from Grandmother's, hoping to cut off her snack supply. But on the one occasion when we both ate supper at the Marin's, I noticed that she ate heartily of everything, although she and Lynn had been snitching biscotti out from under Inez's nose all afternoon. Of course. These were her friends. Here, she was comfortable.

Liz's connections formed a web reaching far beyond the family. If Daddy blocked one strand of the web, others remained to hold her up, to lead her out. The Marins were only one of her alternate families.

When Inez and Mrs. Watson went hunting by themselves and came home with a buck they had shot, gutted, and loaded into the truck all by themselves, Liz was there to watch and admire, because she was visiting the Watson girls at their home on a ranch on Cottonwood Creek. When Hilt's new recreational equestrians gathered on a Saturday morning in front of the store, Liz was there, among friends, and as Audomar DeClerck swung her up behind him on Peaches, his big palomino mare, Liz waved at me. The party trotted away across the railroad tracks, gone for several hours.

I studied books; Liz studied people. She learned how to count change because she hung around the counter at the Company store and watched the clerks do it. She learned to keep a baseball box score by peering over Shirley DeClerck's shoulder at Little League games. She carved out alliances that saved her some of the annoyances of life in an inbred and prejudiced place.

When she joined Brownies, for instance, the troop leader, Mrs. Luper, wrote her last name down in the roll book as "Brannon." We had both been signing our last name that way since Mother's remarriage, and our teachers had accepted the change without comment, but two years before, when I joined Brownies, Mrs. Luper would have none of it.

"*Legally,* it's Johnson," she had said nastily, her gaze sweeping around the room, drawing all eyes to me, while I wondered why a woman with such large tufts of black hair under her arms insisted on wearing sleeveless dresses. And she wrote "Johnson" in the book, and seemed to sneer each time she said it.

When I complained to Mother, thinking that Mrs. Luper was just being nasty about The Divorce, Mother only sighed and said no, that wasn't it; Mrs. Luper had been a Haynes and the Hayneses hated the Roushes and anybody related to the Roushes.

But when Elizabeth joined Brownies, she arrived arm in arm with

Lynn Marin, and Mrs. Luper must have made a quick decision, for with only a little hesitation and a large insincere smile, she wrote the name down in the book exactly as Liz recited it to her. Lynn had many cousins in Hilt, and they were all in the scouting program, and they all had parents. Mrs. Luper could count.

When Liz entered first grade, a little boy with olive skin, curly black hair, and big brown eyes sat in front of her. He said absolutely nothing for the first several days of class. He watched the teacher, but when she told the class to take out their pencils, he did nothing until he saw the other children taking out theirs. Not until the second week of school did Mrs. Rutledge learn that Stevie Avgeris's parents were Greeks who lived north of town on a stump ranch. He milked three goats before he caught the school bus, but he could not speak English.

Liz undertook his education. Tapping him on the shoulder, she would hold up a pencil. "Pencil," she said. The other children watched her and followed suit. By the end of the school year, he spoke fluent English.

We were performers, Liz and I, but Liz's talents fell firmly into the oral tradition. She could tell a tall tale with a straight face, as when she came home and told us that her teacher had found a china cat in her attic, with a large emerald tucked into the dried clay inside. She told the story with such detail and conviction that the amazing coincidence that just such a china cat had been described on a television show only the week before temporarily escaped our mental fraud meters. I had begun to write stories, but Liz lived them.

When I remember Liz in the days before Daddy, she is often performing. She dances at one of the annual talent shows in the old Clubhouse, wearing a long striped dress, swatting a tambourine, and singing in her almost inaudible soprano about a swarthy gypsy. Or she stands beside me in the same building, singing Christmas carols at the finale of the annual Christmas program, as Santa Claus comes up the aisle. Every child in town comes forward and is handed a mesh bag full of Sunkist oranges, apples, a bag of Hershey's kisses, another bag of hard candy all stuck together, and at the bottom a mound of pecans and walnuts and Brazil nuts and hazelnuts and almonds, tucked around a big pink popcorn ball oozing hardened syrup, which we would never eat.

Liz manages to turn her presentation into an Academy Award performance, reaching up on tiptoe to peck the furiously blushing Santa on one cheek.

During her eleventh summer, she and Johnny Foggiato performed a strange little pas de deux on the sidewalks of Hilt, each for their own purposes.

We first knew Johnny Foggiato as a handsome black-haired high school boy, raised in Little Italy. By the time Liz was eleven, he was going to Stanford on an engineering scholarship. Everyone in town was proud of him. The Company gave him a job tearing up and rebuilding several of the old board sidewalks in town. He started on Front Street. It took Liz approximately two minutes to discover him and form an alliance.

As Johnny, stripped to the waist and tanning beautifully in the summer sun, began pulling up the gray, rotting boards with a crow bar and dropping the big rusty nails into a coffee can, Liz was there to look under the boards. Occasionally, among the blanched dandelion roots and earthworm casts, there were coins. "How about you put the nails in the coffee can for me," Johnny proposed, "and we split any money we find, fifty-fifty?" He had a deal.

When Johnny took a break, Liz ran down to Grandmother's and came back with a jar full of lemonade. Wherever Johnny was that summer, Liz's bike was leaning on a fence nearby, and she was picking up nails as he popped them free of the boards. It didn't occur to me until years later what Liz was really doing for Johnny: running interference.

He was handsome and smart and going places, was our Johnny. And every nubile female in Hilt knew it. Many of them found some excuse to pass the sidewalks where he worked, but Liz was always there, so they could only smile and say hello and keep walking. Johnny was only human—he couldn't resist showing off his muscles and his tan to the local talent—but he also knew where he was going, and it wasn't to a mill job in Hilt. And to get where he was going, he had to stay away from the girls in the tight halters and shorts who followed his progress up one sidewalk and down the other. With Liz there, he was safe, at least during the workday. Young as she was, my sister knew a mutually profitable relationship when she saw it. And of course, she was in love with Johnny, too.

CHAPTER 22

This set of sagas, memory. Over and over self-told,
as if the mind must have a way to pass its time, docket
all the promptings for itself, within its narrow bone cave.
IVAN DOIG

Once I stopped expecting anything from Daddy, my life became, oddly enough, better. With love taken out of the family equation, I could concentrate on the certainty that if I worked hard—if I scrubbed the floors and learned to iron shirts and kept the woodbox full and didn't burn the bacon or overcook the eggs—then there would be no more talk about sending me away. If Daddy wanted to define affection as work and good behavior and being seen and not heard when company was present, I could do that. If I behaved myself, no one paid much attention to me, and as far as I was concerned, that was all to the good.

Now, when Daddy came home, I made sure that he always found me working at something: folding clothes, ironing, bringing in wood. Sometimes I had to throw my copy of *Mad* under the couch cushion and skid into the kitchen when I heard his truck, but working had started to become a habit, something I did automatically, while my mind traveled other paths and came to some strange destinations. For even as I told myself that I didn't need what I couldn't have, I was still trying to earn the love of my stepfather, earn back the love of my mother. At the same time, I grappled with the revelation that marriage did things to women, bad things. Grandfather upset Grandmother.

Mother had been happy and kind before she married Daddy. Marriage took something away from women, so that they had to do what their husbands told them.

I did not understand this, but the solution seemed clear to me: don't get married. Marriage and children were traps for women. But I could avoid them. I could earn my own living someday and live alone and do what I liked.

In the woodshed lean-to, I spent more and more time chopping firewood with Daddy's great double-bitted falling axes, gradually learning the logger's full roundhouse swing. When I came to a big knotty piece, I rummaged around and found the iron wedges and the sledge hammer and pounded the sharp wedges into the wood until the stubborn piece split, with a satisfying crack. I chopped kindling, delicately splitting the pieces of straight-grained, pitchy wood until they shattered into slivers. I learned the sharp pissy smell of green black oak, moss-covered and heavy, as I stacked their broken rounds higher and higher under the eaves.

At school, some of my classmates tried to talk to me about Daddy. "I think your dad's too hard on you," Helen Shepherd said to me one day, in her quiet, gentle voice, as we idled side by side on the playground swings at recess. "My dad never spanks me." I looked at her, at her white skin and auburn hair, at her soft hands and slender wrists. In another year, she would be wearing lipstick and trying out for cheerleader. She couldn't hit worth a darn, and she never had to bring in the wood. When she walked to the store on weekends with her father, he held her hand, and bought her Popsicles.

"Well, I'm glad mine does!" I snapped at her with a sudden flash of temper and was rewarded by her hurt expression. I was immediately sorry, but just then the bell rang, and Helen leapt up and ran for the school as though a devil was after her.

At home, the adults had other things on their mind. Jo's return had complicated life for them, especially for Grandfather. His tense standoff with the Company had resumed; his life had become a long waiting. If he were still employed on January 1, 1962, the Company would have to start writing him a monthly pension check. But if he was "let go" even one day before that, there would be no pension and nothing to look forward to, except collecting Social Security when he turned sixty-five. He would probably not be able, in that case, to afford the one dream for which he had saved all his life: a house of his own. To Grandfather, the house on Front Street was just a place to wait for retirement, just a roof over his head. That was the major difference between us. To his children and his grandchildren, that house was our real home, and it really belonged to Grandmother.

In the summer of 1961, the box factory did not open again after the annual two-week vacation. The equipment was dismantled and cannibalized for use in other parts of the plant, or scrapped. Technically, Grandfather was out of a job, but he continued to go to work—at a desk in the Company office—and to figure costs and crunch numbers. But the people who had regarded him as superfluous six years ago still worked in the Company headquarters, and their opinion of him had not changed. He had to campaign as hard as ever Nixon did, pointing out to Mel Barron that the Company would actually be saving money by keeping him on for another six months. He had come to believe that he and the Company had struck a deal. Indeed, Daddy said the scuttlebutt around the office was that the Company had already decided that it would be cheaper to do just that; if they fired Grandfather before his pension was vested, he would hire a lawyer and sue them, and they knew it. Though the Company would probably win such a lawsuit in the end, it was cheaper to give him what he wanted in the first place. And any insurance underwriter could tell you that most men Grandfather's age were dead in five years, anyway.

Within a few weeks of the day that Mother brought Jo home, rumors started about her. When I asked Mother about them, she wouldn't tell me, and I assumed that the rumors were only about Jo's divorce. Still, Jo had become a Problem once again, and I heard Daddy talking to Mother about her. "She just won't grow up and take responsibility for her life," he said. "She doesn't have any self-discipline." And Mother, sorrowfully, could only agree. I remembered Jo's apartment, where piles of wrinkled clothes engulfed the sofa and the beds were unmade and the dirty dishes piled up on every surface. Perhaps they were right about her.

One day the telephone rang as Jo sat drinking coffee with Mother in our kitchen. When I answered it, a woman's voice asked if Josephine was there.

"Josephine?" I repeated. Nobody called her that.

Jo's arms began semaphoring wildly, and she shook her head violently from side to side.

"No, I haven't seen her all afternoon," I said, feeling very grown-up as I replaced the receiver. Jo pantomimed pulling hanks of hair out of her head.

"Who was that?" Mother asked me.

"She didn't say, but it sounded kind of like Mrs. Chase," I said, puzzled about why Jo wouldn't want to talk to her.

"She just wants to know where I am," Jo said.

Mother stood up and got her purse from her bedroom, pulling out a dollar bill. "I need a quart of milk and a can of creamed corn," she said, holding the money out to me. I hung up the dishtowel and leaped for the door. I loved running errands. As I slapped open the outer screen door, I heard Mother's voice lowered in seriousness to Jo.

"What in the world is going on with you?" I heard her say. I paused, listening.

"I'd better hear that gate open and shut in a minute!" Mother yelled in what we called her fishwife voice, and I gave up and headed down the alley.

None of Jo's old girlfriends lived in Hilt now, and I remembered the old times when a walk around town with Jo would bring Rosalie and Candace out to lean on the fences and talk. Now we walked alone with her, out toward the SS Bar Ranch, Jo pushing Sidney in his stroller over the lumpy chip seal. Sometimes when we came back, we found Grandmother in the front yard, scanning the horizon anxiously. "What are you so worried about?" Jo said irritably. "We just went for a walk." I couldn't understand why Grandmother should be so worried about Jo, in Hilt of all places.

About this time, Grandfather became more interested in his ancestors. He sent for a book of Roush family history, and he marked the page that gave his own name and his father's. I looked at the book, too. I learned that Grandfather's paternal great-grandfather, Permenias Roush, was born in 1802 in Ohio, and that he married a Catherine Smith, who had nine children with him. His third son, John, married an eighteen-year-old Kentuckian, Sallie Ann Scott, in 1858.

So here were the Scotts, who Grandfather had once described as moonshiners and outlaws. Here was Sallie Ann, perhaps the best of a bad lot. She had died, I saw, two years after her marriage, leaving a son, also named Permenias. Sallie Ann, then, had been Grandfather's grandmother. Pointing to her name in the book, I asked Grandfather about her one rainy Sunday afternoon. What did she look like? He put down his newspaper and stared at me.

"My father," he said at last, "had an old photograph of her, what they called a daguerreotype. It wasn't a very good picture."

"Was she pretty?" I asked.

The newspaper started to come up again. "Yes, she must have been quite pretty," he said, his eyes now out of sight. "She looked a lot like your Aunt Jo."

Jo remained in Hilt through the autumn, and through Thanksgiving, and Christmas, and was still there on Retirement Day, when Grandfather drank three pitchers of martinis and almost passed out in the middle of Grandmother's traditional New Year's Day dinner of leftover-turkey tamales. In the months to come, she and Sidney would stay home while Grandmother and Grandfather took house-hunting trips to central and southern California, looking for a place to retire. After they came back from one of these trips, Grandfather accused Jo of precisely what the gossips were saying, of having an affair with Lester Chase. Jo snapped back that as long as he was bound and determined to think the worst of her anyway, he might as well know that while they were gone Lester had come in through the back door, cornered her in the pantry, and raped her.

"Was it true?" I asked Mother many years later.

"I wish I knew," Mother said. "Even Grandmother wasn't sure if she was telling the truth. And of course Grandfather would never have allowed her to press charges. He would have been too afraid of what Lester would say in court."

Late in May of 1962, Uncle Carl came to Hilt to visit his son. He and Jo got into his car and went for a drive with Sidney. When they came back a couple of hours later, Jo emerged from the car, smiling, and trotted up the front steps. She took off her diaphanous scarf and shook out her curly hair.

Liz and I were sitting on the front porch. "Congratulate me, girls," she cooed, holding out her left hand to us. "I'm engaged!"

Carl picked us up and hugged us the way he used to and drove off, waving. He was back a week later, and he and Jo walked over to Bob Trinca's house and were married by him in his capacity of justice of the peace. And as quickly as that, Jo and Sidney and the crib and stroller were gone.

A week later, Grandmother began packing for the move to the new subdivision near Oroville, California, where she and Grandfather were buying a brand new house. But for Carl's intervention, Jo would have had no choice but to go with them and to live in a house in which Grandfather would be home all day, every day. Marriage might indeed be a trap, but it could also be an escape from something worse.

CHAPTER 23

Only the rocks live forever.
NATIVE AMERICAN PROVERB

And then the universe changed. Daddy became Dad. Tommy stopped throwing up. Liz and I carried him atop our shoulders all around the yard and down the sidewalks. Once Liz tripped and fell while carrying him, after he had demanded more and more speed. He shot over her shoulders into the gravel and took all the hide off his chin. We soothed him and bandaged his chin and took him back outside with us.

One Sunday when Liz and I were practicing softball, he ran behind her just as I pitched the ball, and I saw the bat in her hands pull back and strike him on the forehead. His mouth opened and he began screaming. I picked him up and raced into the house with him, watching as he fell silent, his breath held. He turned from red to purple to blue and passed out. Mother and Dad held him over the kitchen sink and threw water on his face until he breathed again, screaming louder than ever. Liz and I didn't mind very much when Dad, his arm almost flaccid with relief, spanked both of us, and sent us to bed without supper.

Late in the seventh grade, a month before my thirteenth birthday, my waist suddenly began to grow in and my breasts and hips to grow out. I found myself actually looking forward to dances in the school lunchroom, where we twisted to Chubby Checker in dim light beneath strands of red and white crepe paper. I wore my first nylon stockings and a garter belt and high heels about a size too small.

We watched our chaperones trying to waltz to "Teen Angel" and smirked with adolescent superiority. Blake Green danced down the hall and tapped the fire alarm box lightly, playfully, with an empty French-fry basket, and the glass fell out and the alarm went off. The building emptied, and a local legend was born.

I grew taller and almost thin. Despite all my good intentions of a manless life, I developed a crush on a boy from Hornbrook, who sometimes came to our dances. I wrote to him. He did not write back.

Liz shot up almost to my height and began to eat more. Dad put up a basketball hoop for us on the outside of the woodshed wall, and we both played on the girls' team at school, where Mr. Rhodes, principal and coach, initiated us into the mysteries of zone defense.

My nearsightedness stabilized, and, despite the basketball, I didn't break my glasses again. I continued to chop wood and to play baseball and Mr. Rhodes timed me at the fifty-yard dash and announced that I should run track.

We rode one night with twenty other kids in a yellow school bus as Mr. Rhodes drove us home in triumph from a tournament, singing our school yell: "We are the Hilt Jets, mighty mighty Hilt Jets, everywhere we go-oh, people want to know-oh, who we arrar, so we tell 'em, we are the Hilt Jets, mighty mighty Hilt Jets . . ."

When Mr. Rhodes saw me collecting rocks and trying to identify them, he found books on geology for me. When we wrote stories in class, he told me that mine were good and saved them and encouraged me to write more.

Liz and I still played most of the old games. We flew kites from the top of Watertank Hill, and when they crashed into the schoolyard, we patiently wound a hundred yards of string back onto a stick. We lay on our backs and watched nighthawks, their wings banded black and white, careening against the evening sky.

The hair on my long new legs turned dark, so I picked up Dad's razor and peeled a long strip of hide off my shinbone and bled all over the bathtub. Mother bought me a Lady Remington electric shaver.

There were few spankings, now. Dad was more likely to withhold our newly awarded allowance of a dollar a week, or order us to copy out sentences like, "I will not stack the good china under the frying pan," five hundred times, or forbid us some long-desired movie or school outing.

One summer day we discovered a house behind the school where a young red-tailed hawk hopped around the back yard, a swivel leash on one foot tying him to a stump. So we met Johnny and Larry, fourteen and twelve years old, and their hawk, Cactus, raised from a nestling. Johnny brought out a heavy leather glove, and Cactus wrapped his wicked talons around our forearms and nibbled at our hair and ate grasshoppers from our fingers. We hiked the hills around Hilt with these new friends, taking turns carrying the hawk, and discussing how best to teach him to fly and hunt, while the bird glared at the world with bright eyes and seemed happy just to be carried through life. On our hikes we killed a rattlesnake and found ruined cabins and a huge iron cauldron in a gulch, and Liz caught poison oak.

Johnny and Larry had been in Hilt so short a time that they carried none of its baggage. They could not have cared less that Dad was really our stepfather. Where we come from, they said, nobody cares about stuff like that. And suddenly, Hilt seemed too small for us, and we ached for those larger places where nobody cared, and when Mother and Dad drove around the new neighborhoods going up on the hillsides around Yreka, I leaned forward, eagerly, and dreamed of streaking through the crystalline blue waters of the big new community swimming pool. The world seemed to open up before me and to hold within it endless possibilities for creating the future—and forgetting the past.

California's game laws allowed children of twelve to get a hunting license if they had completed an approved hunter's safety course. Every year, Bill Straight taught such a class at Hilt's elementary school. It met twice a week for several weeks and was capped by a Saturday outing, which involved the expenditure of live ammunition at targets. I signed up for the course in August of the year I turned thirteen.

As Warden Straight wrote down my name, he looked at me with his head tilted, squinting out from behind his black-framed bifocals. I got the impression that he thought I had done something wrong. I asked Dad—we always called him Dad, now—about this later that evening, and he looked up from his easy chair and replied with a question.

"Did you know that it's against the law to have venison in your freezer unless it's stamped by a state game warden?" he asked.

I pictured the venison in our big home freezer, all unstamped. "But we don't . . . then why?"

"Because if the game warden had to stamp everybody's deer meat, he wouldn't have time to do anything else. So it's not enforced. But if they suspect someone's poaching, it's a good thing to get them on if you can't get them on anything else." Dad looked at me, steadily, for a moment, then sighed and shook his head, but not at me.

"Bill Straight thinks I shoot deer out of season," Dad said. "And he'll take that out on you. But if you want to go hunting, you'll just have to keep your mouth shut and get through the course. Can you do that?" I nodded, and he snapped his magazine back up and began reading about a man who shot turkeys in North Carolina.

So I learned that there was the law, and there was justice, and they were not necessarily the same, in or out of our home. It made it easier to understand why Dad was having trouble with Doug Whittaker, his boss.

By 1962, Dad figured the Company could mill logs at Hilt for another ten years. After that, he told us, they wouldn't have enough large timber left to support the sawmill, although they could still sell stumpage. Unless they could buy timber from adjacent National Forest lands to make up the difference, the mill would have to close.

The Company was already bidding on National Forest timber sales, and Dad had more and more contact with Forest Service employees, and the more he saw of the Forest Service, the better it looked.

Doug Whittaker had worked in Westwood before coming to Hilt in 1958, but he had never run a woods operation before. He was a lay minister for his church's outpost in Yreka. Doug looked at Dad and saw a nice young man with a nice young family, a prime candidate for proselytization. Dad looked at Doug and saw the poster child for everything he had always hated about organized, hierarchical religions. When Dad finally lost patience with Doug's repeated sales pitches and told him curtly that he wasn't about to join *any* church, much less Doug's, the honeymoon was over. Dad found himself working the weekend fire watch, three and four weeks in a row.

At about the same time, Fruit Growers' loggers began to notice that since Doug arrived, most of the new hires were members of Doug's church. The next time the representative for the loggers' union came around and canvassed for an election, the loggers voted yes. Fruit Growers woke up one morning with a closed shop in the woods. The Company was not amused, but when it didn't fire Doug, Dad sent in his application to the Forest Service. When the job offer came through, he gave a month's notice and wrote a letter telling the Company exactly why he was leaving.

The new job with the Klamath National Forest meant a temporary drop in salary, but a real career ladder. Mother and Dad drove to Happy Camp, a logging town seventy miles from Hilt, on the Klamath River, to look around. The following week, Mother drove back by herself to find a house to rent. She visited Alice Dunaway, an old friend of her Aunt Frances. Alice ran a rooming house in Happy Camp; she told Mother to talk to a Mrs. Fowler up Indian Creek, who showed her a fairly new small house with a detached laundry room that rented for eighty dollars a month.

In October of 1962, we drove out of Hilt for the last time as residents, and the valley and the town disappeared behind the rise of Bailey Hill, behind the slope of Sheldon Rock. We looked ahead to the big timber and the wild country of the middle Klamath River, and we were not unhappy.

Braced between my feet, Tommy's pet sun-perch sloshed in his goldfish bowl. The long-wished-for puppy stood on Tommy's lap, head hanging out the window. In the front seat, Liz unfolded a map and looked ahead to the unknown lands. Mother sat atop her driving pillow, clutching the Ford's steering wheel, eyes fixed straight ahead. Before us, Dad drove the blue International pickup, successor to the old red Chevy, loaded with the contents of his workshop. And ahead of us all, Uncle Charlie steered a big U-Haul truck, cigar between his teeth, his grin showing in the rearview mirror.

In 1966, Dad inspected a Forest Service timber sale on the National Forest, some fifty miles from Hilt. He walked into the logging unit and found Frank Benson, our old neighbor, dragging a turn of logs down a creek. Perhaps Dad would have derived some satisfaction from knowing that Doug would read his signature on the violation notice, if he hadn't already known that his old boss had been consigned to a cubbyhole, with no authority over anyone, from the day that Mel Barron read Dad's letter of resignation.

Dad and Frank had a good visit. Times were good at Fruit Growers. The expansion of the Vietnam War the year before had turned up the heat on the lumber market, and the Company had made changes, adding a band saw to cut smaller diameter logs and putting a barker in the sawmill and a chipper in the planing mill. These were just survival tactics, Frank said; Fruit Growers was still Fruit Growers. Replacing the old fuel house with an open sawdust pile, for instance, was just plain dumb. The pile got wet in winter and made it harder to operate the steam plant.

The Company still wanted every division to make a profit, so the woods department had to buy cedar culverts from the carpentry shop, and when those became too expensive, the woods boss quietly bought galvanized culverts from the outside, and the wooden culverts were eventually thrown into the burner. The sawmill and the rest of the plant still worked a single day shift, and the entire plant still shut down for the two-week vacation in August. Only the truck shop stayed open to perform heavy maintenance on the logging equipment.

As the Company prepared to log Jaynes Canyon, one of the last roadless drainages in the Beaver Creek watershed, Hilt slowly emptied. More and more people were buying houses in Yreka or Ashland and commuting to work. Several houses had been torn down when it became obvious there was little demand for them. The modernization of the mill was expensive, and the Company needed cash. We could see this for ourselves. Sometimes, as we drove through Hilt on our way back from Medford in the summer, we noticed how shabby the little town had become. But by then Liz and I missed it, terribly, and we rolled down the windows and drank it all in, trying to hold on to it one more time.

By 1970, the demand for lumber was so high that white fir logs less than twenty inches in diameter were marketable. The Company upgraded the mill once more to handle them. "I'll bet," Dad said, "that they wish they'd bought the land that Southern Pacific wanted to get rid of in the late fifties."

"Why didn't they?" I asked.

"It was high-elevation white fir—not worth milling then—and Fruit Growers always thought they could get plenty of National Forest timber, like they did in the thirties, if they ever needed it again. They never thought there'd come a day when everybody else would he trying to buy government timber, too."

Southern Pacific had started building roads on its lands near Hilt in the 1950s, when they harvested old growth near Sterling Mountain. By the late 1960s, they were logging in Dead Cow Creek, Jaynes Canyon, Dry Lake, Wards Gap, and Millers Glade, all high-elevation white fir and red fir sites. By the 1970s, they had moved on to Trapper Creek and Fat Doe Creek, building roads up the side drainages, until the old growth was gone. Southern Pacific sold its Beaver Creek holdings in the 1980s, and their successors today own several thousand acres in the watershed.

The end came quickly for Hilt, after all. In October of 1972, the board of directors of Fruit Growers Supply Company met at their headquarters in Sherman Oaks. When they sat down, Hilt was a community. When they stood up, it was a ghost. The Company president came to Hilt and told the employees and issued a statement to the newspapers. "Today's costs of operation make it impractical to continue this operation on its present one-shift basis. To increase this operation to two shifts would result in depleting Company-owned timber reserves in five years or less. This is not ecologically or economically sound."

Audomar DeClerck, the green lumber foreman who had worked for Fruit Growers for thirty years, and his father for fifty, told a reporter, "It's like getting news that the breadwinner of a family is suffering from a terminal illness and has less than a year to live." The Company would keep only a forestry crew of about ten people, who would work out of the Hilt office and manage the timber, the way the Company's lands near Susanville and Westwood were managed.

In November, the 125 loggers went home for good. The 165 sawmill workers kept up the regular day shift until June 1973, when the decks were exhausted and the last log went through the mill. I was not there, but in the yellowing newspaper clippings I see Hilt as it must have been that day late in June, when the last log cleared the carriage of No. 2 rig in the afternoon and was sent up the slip by Ray Middleton, as his brother Rich cut the top sixteen-foot section on No. 1 headrig, and his brother Darryl cut the eighteen-foot butt section on No. 2 rig.

Left in the woods were 550 million board feet of standing timber. In sixty-three years, the mill at Hilt had cut more than a billion and a half board feet of timber. The foresters estimated that the Company's forestlands grew fourteen or fifteen million board feet per year, so by selling twenty million per year, the stumpage should last another one hundred years. "Simple arithmetic," said a Company memo, "makes it obvious that such a program will provide the maximum return to the citrus growers who are the owners of Fruit Growers." Simple arithmetic had made Hilt, once, and now had unmade it again.

Prices for incense cedar stumpage soared in 1974, so the Company sold a million board feet of pure cedar in a single sale, from an area where all the pine and fir had already been cut. The Company also managed to sell fourteen million board feet of other species that year, as well as four million of dead and down cull logs (suddenly valuable in a soaring woodchip market) left on the ground from previous cuttings. On the cleared land, the Company silviculturist, Nick Freemyers, supervised the planting of conifer seedlings.

In the 1990s, the Klamath National Forest began gathering data and writing ecological studies on the forest's major watersheds. I sent for them and read them, and as I waded through the turgid bureaucratic prose, the gray winter landscape of eastern Idaho fell away, and I felt my own country come alive for me again and began to understand what had happened to Hilt and to all of us.

When its lands in the Beaver Creek drainage played out, Hilt disappeared as a town. The young trees growing on the old skid trails, and coming slowly up through the brush in the old burns, would never see the inside of the old sawmill. Long before they would make a two-by-four, the mill would be closed. Yet many in Hilt never saw it coming.

We had all been tied to the land and the fish and the frogs and the trees, without really knowing what that meant. That something was wrong in our ravenous consumption of trees and building of roads had long been camouflaged by the very continuity of life in Hilt. The logging trucks rumbled into the mill, and the lumber came out, and there was no end to it. Smoke rose from a hundred stovepipes and from the mill burner and hung low in the valley in the winter mornings. We thought it was forever, because it was all we had ever known.

As I read through the thick Forest Service studies, I felt a cold sadness in my stomach. For the Forest Service had thought that its timber program was forever, too, and only now, when it was too late for all the lumber mills that had survived Hilt, was it speaking the truth that a few Forest Service people had always known and had been smart enough not to talk about.

For over a hundred years the trees fell in Beaver Creek, cut down by loggers and miners and road builders. Roads crept up the bottoms of all the streams in all the branches of the drainage: up Hungry Creek and Grouse Creek and the West Fork. The road cuts and fills eroded, sending sand and silt into the creeks, filling up the spawning beds, destroying the pools, choking the eggs of salmon and steelhead.

When the trees were gone, the shade was gone, and no old trees were left to fall into the water and form sheltering pools. The beaver did not come back, this time. The living creek was now only a mined-out channel, with no willows, no alders, no food.

Once, ponderosa pine and sugar pine grew in open stands on the southern and western sides of the slopes, and every ten to twenty-five years, on average, a fire burned through, thinning out the young trees and cleaning up the fine fuels like needles and twigs. Because of their thick bark, the large trees easily survived these ground fires. Below three thousand feet, hardwoods blended into the mixed conifers. On northern and eastern slopes, Douglas-fir predominated, and fires kept the encroaching white and red firs at bay. Above five thousand feet, red firs grew in a mosaic with wet meadows.

By 1934, almost all the big sugar and ponderosa pines in the eastern half of the Beaver Creek watershed were gone, plucked out by railroad and steam-donkey logging. Twenty years later, the Douglas-firs went, and in the 1960s the white firs and red firs and incense cedars were cut, as the logging moved higher and higher. The smaller trees, the younger ones, were left. But young white and red fir trees are easily skinned and damaged in logging operations, and the wounds become infected with rot. They also burn far more easily than pines.

As successive logging entries opened up the canopy, and as the Forest Service successfully promulgated its mantra of fire suppression and prevention, the drainage continued to change. Dense stands of young Douglas-firs and white firs now grew on the south-facing slopes in place of the old ponderosa pine stands. As fire was controlled, they survived, in numbers unthinkable before.

Between 1922 and 1994, the Forest Service documented 473 fire starts in the Beaver Creek watershed. Most of these were lightning fires. All told, they burned 17,500 acres, including areas burned more than once.

The 1912 Forest Service timber reconnaissance found over 400 million board feet of timber in the Beaver Creek watershed—245 million on private lands, over 165 million on National Forest lands. In the years since, the Forest Service had sold more than 112 million board feet of Beaver Creek stumpage.

Most of the timber in the Beaver Creek drainage was actually cut beginning in the 1960s, when clear-cutting really arrived on the Klamath National Forest. The last government sales in the drainage were negotiated in the early 1990s, just as logging in spotted owl habitat came to a virtual halt all over the National Forests in the Pacific Northwest. In all of the Beaver Creek watershed, only about fifty-five million feet of old growth remained on the National Forest sections. When Dad took that first Forest Service job in Happy Camp in 1962, the Happy Camp Ranger District was cutting fifty-five million feet every year, enough to supply the needs of Happy Camp's two largest mills.

The Forest Service now plans to sell only a million board feet a year from the Beaver Creek watershed. State regulations, which limit cutting in spotted owl habitat, have lowered the amount of timber that Fruit Growers and other private owners can sell, too.

But this is not the whole story. "Most of the big trees on National Forest lands in the watershed were cut a long time ago now," a Forest Service officer told me recently. "The lumber interests like to publicly tout the value of logging as a preventative of big stand destroying fires, but in private they tell us that they don't want the small-diameter logs that a real stand-opening operation would harvest. They want the very,

very last of the big trees, and they don't want to bid on sales that don't have them."

Today, fish habitat in Beaver Creek is poor. The granitic soils, laid bare by roads and clear-cuts, continue to bleed into Beaver Creek and its tributaries. With the end of the big sales that brought in the big money, the Klamath National Forest's budget continues to decline. And because so much of the rest of the Forest Service budget has historically been tied to timber, as logging declines, so does money for fire prevention.

In the years to come, the Forest Service now admits, the number of fire lookouts will be reduced; the number of airplane overflights will be cut. Fires will be bigger when first detected, and in the heavy fuels and thick young timber, the fires will burn hotter and eviscerate entire subwatersheds, which in turn will pour more sediment into the creeks.

The Klamath National Forest now has virtually no money for road maintenance; the roadwork now being done in the drainage is done by the private timber companies. Fruit Growers is steadily obliterating and revegetating many of its old roads, thinning a stand, then backing out, doing the work that the Forest Service cannot afford to do. Close a Forest Service road, and current vegetative conditions combined with reductions in fire personnel make a high-intensity fire more likely; leave the road open, and erosion threatens a creek. And the Forest Service has to listen to the public, while Fruit Growers does not.

"Every time we want to close a road," a forester on the Klamath National Forest told me last year, "the Siskiyou Houndsmen raise a stink and we have to go through a big public-involvement process. Fruit Growers doesn't have to do that." I remembered those signs on Fruit Growers' logging roads, long ago: "Private Road: Permission to Pass Over Revocable at Any Time."

The crowded stands of fir began to die in the drought years, when there was simply not enough soil moisture available to support them all. Over 50 percent of the Beaver Creek drainage now has a high potential for extreme fire behavior. Twenty years ago, no one thought that the new and crowded fir stands replacing the old, fire-resistant pine stands were unsustainable. The Forest Service never thought that, by the 1990s, sale offerings in those crowded second-growth stands would go begging. Yet it had created those stands deliberately, believing that by the last decade of the twentieth century, the timber famine that Gifford Pinchot so often predicted would at last have come to pass.

The Forest Service's studies say that only a long-term burning program, to bring back the gentle fires and recreate the open pine forests that the first lumbermen found, stands between the Beaver Creek drainage—between the entire Klamath Mountains—and a future as a gigantic, self-perpetuating brush field. The problem is that the Forest Service is broke. For fifty years it gave Congress what it wanted—timber for powerful constituents—and now that the big timber is gone, Congress has no money for the Forest Service. So the agency waits, helplessly, hoping that the predictions written into its watershed studies will not come to pass, yet absolutely certain that they will.

"The fires will occur," my forester friend says. "The only question is when. These forests evolved in a disturbance-driven environment. And when it happens, it's going to be big."

I can hear the pessimism in his voice, a thousand miles away. "There's a name for what's going to happen," he says. "It's happened before in Mediterranean climates—in Greece, in Turkey, in the Lebanon. California's no different. The truth is that a forest can survive anything except people, because only people keep taking and taking and never stop, never just leave the forest alone and let it be a forest. The only people who ever lived by just picking up the surplus that the forest was capable of giving, and leaving the rest of it alone to support the ecosystem, were the Indians. And look what happened to them."

EPILOGUE

> Many shapes the gods will take,
> And many things they bring to pass
> Beyond our hope, and what we sought
> Is not fulfilled by them; and for
> Undreamed of things God finds a way.
> EURIPIDES

My grandparents lived together in their new house until 1971, when Grandmother died of cancer. As she lost weight and strength, Mother went down to help care for her. When she became bedridden, Grandfather made arrangements to move her into a nursing home. As attendants lifted her into the ambulance, Grandmother looked up at him, and her vacant blue eyes became, for the last time, absolutely lucid. "I know what you're doing," she said. Grandfather went back in the house and cried. Mother went back in the house and had a double highball.

Grandfather lived for four more years. When he died, his body lay undiscovered in the house for three weeks, because by then he and Mother weren't talking much, and Jo was far away.

Dad worked for the Forest Service for eight years, rising to timber management officer before a brain tumor forced him to retire in 1970. He died in 1978.

Mother worked for the Forest Service for ten years after Dad retired. She lives nearby, tends a yard full of herbs and roses, paints watercolors, and tells me at last the stories I never heard.

Jo had another baby, a girl, in 1963 and divorced Carl a year later. In 1966 she married again and moved to Arkansas. She now lives in a small town in Kansas, where she plays the organ for the local Roman Catholic Church.

After a year of college, Liz joined the navy, served two hitches during the Vietnam era, then married a sailor from Illinois. She went to nursing school with her GI money and now works for a large hospital in central Illinois. She has two daughters and recently became a grandmother.

Tommy joined the navy out of high school, served two tours on an aircraft carrier, then worked his way through engineering school. He has done well in his profession and is married with one son.

I went to college, got a degree in English, and took what I thought would be a temporary job with the Forest Service. I've been with the outfit for twenty-eight years now. I am married and have no children.

Dick Johnson came back into our lives in 1991, after his second wife died. "You look exactly the same," he told me, as he stepped out of the car to hug me. He is married to Bette, a small neat woman who is in some ways eerily like Mother.

And we found her, the red-haired child my aunt left behind. Two years ago, I put a notice on an internet site dedicated to reuniting children with their birth mothers, mentioning only birth date and place. Six months later, I got an email from California. Her name was Nancy, and she had been in foster care until she was three months old. The state agency had had a hard time placing her because of her red hair. She sent me her phone number, and when she said hello, I knew beyond doubt that I had found our lost cousin. She sounded exactly like Liz.

"It's funny," she told me. "I always knew I was adopted, even before my mom and dad told me. When I played by myself, I'd call myself by another name, and I sort of knew that it'd been my name once."

"What did you call yourself?" I asked, although I knew the answer.

"Martha," she said.

In her seventh decade, Jo now knows that the child she gave away received love and music lessons and a wonderful childhood with parents who ran a pet shop in a city and adored her. But she still wishes she could dig Grandfather up and kill him again. Sometimes we all do.

A few days after Grandmother died, my sister called me. She had come suddenly awake the night before, to see Grandmother standing at the foot of her bed. "It's all right," Grandmother told her. "Go back to sleep."

"Well," I said, "I had the Hilt dream again last night," and told it to her once more, this recurring dream that has come to me now and then ever since we left Hilt.

In the dream, I stand alone on the streets of Hilt, on a night without stars, in front of the Company store. I run down Front Street toward Grandmother's house, and the houses I pass are dark and deserted, like the street, like the town. I come to Grandmother's house and see the big front room window gaping, black and dead. The old tree's branches thrash in a cold wind, leaves rattling insanely, a stranger. The leprous undersides of the leaves flash under the dim streetlight, just inside the unpainted fence.

But as I approach the house, I see the big white front door opening onto a lighted room. And standing in the doorway, holding the screen door open, is Grandmother. I run up the steps and into her house once more. Around the edges of the front room, the white wainscoting gleams, for the room is stripped of furniture and drapes. Only the big mirror remains, hanging on the southern wall. Grandmother is not surprised to see me, nor I her. She speaks to me, and I understand her perfectly, although later I cannot recall the words. I follow her as she walks through the old house, and somehow I know that she has come back here to stay, to wait for Mother and Jo and Liz and me. And I wake up with her voice in my mind and carry with me, for a whole day, the certainty of her continued existence.

Our growing up is beyond our control, dependent on others for the good or evil of place and people. We are blessed where we do not deserve, cursed where we have done no harm. In Hilt we had security and certainty beyond what is granted to most people, and we thought it was forever. So when John Brannon came into our lives and changed everything, it was our first hint that the world was larger than we had dreamed and contained—we soon discovered—more than we really wanted to know.

After he died, for a long time the four of us who were left did not talk much about him. We handled the knives and the obsidian projectile points he had made and oiled the guns he had shot and left those first years alone.

Mother gave me his old hard hat, covered with flaking orange paint, and his work shirts, and I wore them to work in my Forest Service job. But most of the things he left us were intangible: skills like shooting and reloading; ways of going and seeing in the forest; silence and vigilance in strange places; watching the river to know where to toss a fishing lure. We judged people by his yardstick, and if we seldom measured up ourselves, we knew that most other people didn't, either.

For Dad was wrong. Love cannot be earned, or deserved, or worked for, or won. But I only learned this years after we had poured Dad's ashes out onto a stump high above the Klamath River, too late to tell him.

What Dad knew, he had learned from his grandfather, the boy who walked to Oregon: to leave before the crunch comes, to live within your means, to recognize limits, to know when to move on. He carried these lessons into his life, from farm child to gyppo logger to private forester to Forest Service officer. We absorbed them from him, and we followed them, mostly. We left the woods behind and left others to learn firsthand what Dad had already known: that the Western dream was unsustainable. Only I stayed longer than the rest, long enough to see the beginning of the end. But that is another story.

The valley where we met John Brannon still holds my heart in its hand, in the curving hills, in the warm ledges of rock and the enduring flow of water and wind that carved it all. Not a day goes by that a trick of the light—afternoon sunshine on old boards, or starlight above a snowy hill—does not bring it all back, and bring it back whole, that cupping of space and time that for just a moment upheld us, until the fingers opened and let us fall.

On the rise where our last house stood, clumps of black locust remain, stunted clones of the tree where our pet jay sang. They drink the rain of heaven and are killed by drought and broken by storms and frost. The roots survive, coming back, pushing upward again, starting over, though the dead tops rattle in the hot summer winds that spill down from Bear Canyon, though the lawns and the flowers are gone and the people—so many of them—dead.

Now in the winter nights, jackrabbits leave their tracks in the vanished yards and coyotes watch behind the sagebrush and cry in the nights and survive. New fences border the old county road, and behind them cattle graze in the spring. Where the boys played baseball, sprinklers hiss and click over a field of alfalfa. On the hill where Italian immigrants raised gardens and children, a small band of elk comes in the winter, to bed down in the oak thickets and watch the winter moons rise over Pilot Rock.

Hilt was the last place any of us would live where we knew with certainty what the next day, the next week, the next year, would bring and where we knew we belonged. In Hilt, the world was graspable and knowable and safe. But Hilt was two worlds, and one of those was about limits and arithmetic. By the time the old-growth forests were gone and the first world ended, the second—the world the children made and loved and still remember—was gone, too.

Bibliography

The following publications were useful in researching the background for this book:

Atwood, Kay. *Jackson County Conversations.* Medford OR: Jackson County Intermediate Education District, 1976.

Beaver Creek Ecosystem Analysis. Yreka CA: Klamath National Forest, 1996.

Collins, William, and Bruce Levene. *Black Bart: The True Story of the West's Most Famous Stagecoach Robber.* Mendocino CA: Pacific Transcriptions, 1992.

Davies, Gilbert W., and Florice M. Frank, eds. *Memorable Forest Fires: 200 Stories by U.S. Forest Service Retirees.* Hat Creek CA: HiStory ink Books, 1995.

---. *Memories from the Land of Siskiyou: Past Lives and Times in Siskiyou County.* Hat Creek CA: HiStory ink Books, 1993.

---. *Stories of the Klamath National Forest: The First Fifty Years.* Hat Creek CA: HiStory ink Books, 1992.

Doron, William, ed. *Chronological History of the Klamath National Forest: The 1950s.* Yreka CA: Klamath National Forest, 1996.

Kroeber, A. L. *Handbook of the Indians of California.* 1925. Reprint, New York: Dover Publications, 1976.

Okrent, Daniel, and Harris Lewine, eds. *The Ultimate Baseball Book.* Boston: Houghton Mifflin, 1979.

Purdy, Tim I. *Fruit Growers Supply Company: A History of the Northern California Operations.* Susanville CA: Lahontan Images, 2000.

Ritter, Lawrence. *The Glory of Their Times.* New York: Collier, 1966.

Roush, Rev. Lester LeRoy. *History of the Roush Family in America.* Strasburg VA: Shenandoah Publishing House, 1928.

The Siskiyou Pioneer. Vol. 4, no. 8 (Hilt Issue). Yreka CA: Siskiyou County Historical Society, 1975.